GCSE Mathematics
AQA
Modular

Sue Chandler   Ewart Smith

Foundation

Module 3

Series editor: Harry Smith
Consultant examiner: David Pritchard

www.heinemann.co.uk
✓ Free online support
✓ Useful weblinks
✓ 24 hour online ordering

01865 888058

Heinemann
Inspiring generations

Heinemann Educational Publishers
Halley Court, Jordan Hill, Oxford OX2 8EJ
Part of Harcourt Education

Heinneman is the registered trademark of Harcourt Education Limited

© Text Sue Chandler, Ewart Smith

First published 2006

10 09 08 07 06
10 9 8 7 6 5 4 3 2 1

British Library Cataloguing in Publication Data is available from the British Library on
request.

10-digit ISBN: 0 435807 21 8
13-digit ISBN: 978 0 435807 21 4

The right of Sue Chandler and Ewart Smith to be identified as joint authors of this book has
been asserted by them in accordance with the Copyright, Designs and Patents Act 1988.

Edited by Carol Harris
Designed by Wooden Ark Studios
Typeset by Tech-Set Ltd, Gateshead, Tyne and Wear

Original illustrations © Harcourt Education Limited, 2006
Illustrated by Phil Garner
Cover design by mccdesign
Printed in the United Kingdom by Scotprint

Cover photo: © Digital Vision
Consultant examiner: David Pritchard, Andy Darbourne
Series editor: Harry Smith

Acknowledgements
Harcourt Education Ltd would like to thank those schools who helped in the development
and trialling of this course.

The author and publisher would like to thank the following individuals and organisations
for permission to reproduce photographs:

Corbis pp **15**, **24**; Harcourt Education Ltd/Debbie Rowe pp **17**, **115**; Photolibrary pp **22**, **27**;
Getty Images/PhotoDisc pp **25**, **144**; Alamy Images pp **59**; Digital Vision pp **63**;
Getty Images pp **70**; Photos.com pp **77**, **127**;
Nature Picture Library/Terry Andrewartha pp **84**; iStockPhoto pp **106**, **141**;
Alvey & Towers pp **130**; Rex Features pp **136**.

Every effort has been made to contact copyright holders of material reproduced in this
book. Any omissions will be rectified in subsequent printings if notice is given to the
publishers.

Publishing team

Editorial	Design/Production	Picture research
Sarah Flockhart	Christopher Howson	Chrissie Martin
Maggie Rumble	Phil Leafe	
Joanna Shock	Helen McCreath	

There are links to relevant websites in this book. In order to ensure that the links are up-to-
date, that the links work, and that sites are not inadvertently linked to sites that could be
considered offensive, we have made the links available on the Heinemann website at
www.heinemann.co.uk/hotlinks. When you access the site, the express code is 7218P.

Tel: 01865 888058 www.heinemann.co.uk www.tigermaths.co.uk

How to use this book

This book is designed to give you the best possible preparation for your AQA GCSE Module 3 Examination. The authors are experienced writers of successful school mathematics textbooks and each book has been exactly tailored to your GCSE maths specification.

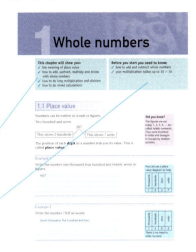

Finding your way around

To help you find your way around when you are studying and revising use the
- **contents list** – this gives a detailed breakdown of the topics covered in each chapter
- **list of objectives** at the start of each chapter – this tells you what you will learn in the chapter
- **list of prerequisite knowledge** at the start of each chapter – this tells you what you need to know before starting the chapter
- **index** – on page 166 – you can use this to find any topic covered in this book.

Remembering key facts

At the end of each chapter you will find
- **a summary of key points** – this lists the key facts and techniques covered in the chapter
- **grade descriptions** – these tell you which techniques and skills most students need to be able to use to achieve each exam grade
- **a glossary** – this gives the definitions of the mathematical words used in the chapter.

Exercises and practice papers

- **Worked examples** show you exactly how to answer exam questions.
- **Tips and hints** highlight key techniques and explain the reasons behind the answers.
- **Exam practice** questions work from the basics up to exam level. Hints and tips help you achieve your highest possible grade.
- The icon **A01** against a question is Using and Applying mathematics. These questions ask you to give reasons for your answers.
- **An examination practice paper** on page 148 helps you prepare for your written examination.
- **Answers** for all the questions are included at the end of the book.

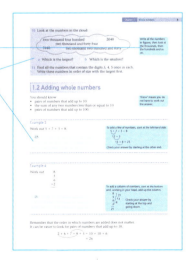

Coursework, communication and technology

- **Mini coursework tasks** throughout the book will help you practice the skills needed for your GCSE coursework tasks.
- **ICT tasks** will highlight opportunities to use computer programs and the Internet to help your understanding of mathematical topics.
- **Class discussion** sections allow you to talk about problems and what techniques you might use to solve them.

Contents

MODULE 3 Foundation tier

1 Whole numbers

Place value	1
Adding whole numbers	3
Adding larger numbers	4
Subtraction	6
Mixed calculations	8
Word problems	9
Multiplying by numbers less than ten	10
Multiplication by 10, 100, 1000, …	11
Long multiplication	13
Division by a whole number less than ten	14
Remainders	15
Division by 10, 100, 1000, …	16
Long division	17
Mixed operations	18
Brackets	19

2 Negative numbers

Negative numbers	21
Addition and subtraction	23
Multiplication and division of directed numbers	26

3 Types of numbers

Names for number	29
Factors	31
Multiples of a number	32
Common factors	33
Common multiples	35
Index notation	37
Square and cube roots	39
Multiplying and dividing numbers written in index notation	41
Prime factors	42

4 Fractions

Fractions	45
Equivalent fractions	47
Comparing fractions	48
Mixed numbers and improper fractions	49
Finding fractions	50
Addition and subtraction	52
Adding and subtracting with mixed numbers	53
Multiplication	54
Multiplying whole numbers and mixed numbers	55
Reciprocals	55
Division by a fraction	56
Fractions of a quantity	57

5 Decimals

Decimal places	61
Changing decimals to fractions	63
Addition and subtraction of decimals	64
Multiplying and dividing by 10, 100, 1000, ...	66
Multiplication and division of decimals	67
Bills	69
Changing fractions to decimals	71
Standard form	72
Using π in exact calculations	73

6 Approximations and estimation

Rounding numbers to the nearest 1, 10, 100, ...	75
Rounding to a given number of decimal places	77
Rounding answers	79
Significant figures	81
Estimating answers to calculations	82
Using a calculator	84

7 Measures 1

Metric units of length	88
Imperial units of length	90
Metric units of mass	92
Imperial units of mass	94
Metric units of capacity	95
Imperial units of capacity	97
Best buys	98
The range in which a rounded number lies	100
Accuracy of answers	101

8 Percentages

Changing percentages to fractions   105
Changing percentages to decimals   106
Changing decimals and fractions to percentages   106
Finding a percentage of a quantity   109
Using a calculator   111
Interest   112
Finding one quantity as a percentage of another 114
Percentage increase and decrease   116

9 Ratio and proportion

Ratio   120
Using ratios   122
Proportion   125
Using proportion   127
Division in a given ratio 128

10 Measures 2

Time   133
Temperature   138
Speed   140
Exchange rates 142

Examination practice paper   148

Answers   153

Index   166

1 Whole numbers

This chapter will show you:
- ✓ the meaning of place value
- ✓ how to add, subtract, multiply and divide with whole numbers
- ✓ how to do long multiplication and division
- ✓ how to do mixed calculations

Before you start you need to know:
- ✓ number pairs that add up to 10
- ✓ number pairs that add up to 100
- ✓ your multiplication tables up to 10 × 10

1.1 Place value

Numbers can be written in words or figures.

Two hundred and seven   207

This shows 2 hundreds.   This shows 7 units.

The position of each **digit** in a number tells you its value. This is called **place value**.

> **Did you know**
>
> that the figures we use today 1, 2, 3, 4, ... are called Arabic numerals? They were invented in India and brought to Europe by Arabian scholars.

Example 1

Write the number one thousand four hundred and twenty-seven in figures.

1427

You can use a place value diagram to help.

thousands	hundreds	tens	units
1	4	2	7

Example 2

Write the number 7502 in words.

Seven thousand, five hundred and two.

thousands	hundreds	tens	units
7	5	0	2

There is no need to write 'no tens'.

Exam practice 1A

1 Write these numbers in figures.
   a Sixty-three.
   b Forty-nine.
   c Seven hundred and seven.
   d Three hundred and twenty-seven.
   e Eight hundred and nineteen.
   f Eight thousand and eight.
   g Six thousand and sixty-seven.
   h Fifteen thousand, two hundred and thirty-four.

2 Write these numbers in words.
   a 56   b 79   c 409   d 187   e 734
   f 330   g 426   h 9488   i 6593   j 7065

3 a Look at the number 96 538.
      Write down the digit that gives
         i the number of hundreds
        ii the number of thousands.
   b Look at the number 70 869.
      Write down the digit that gives
         i the number of units
        ii the number of thousands
       iii the number of hundreds.

4 Write these numbers in order with the smallest first.
   a 55, 43, 61, 57   b 83, 31, 49, 27
   c 308, 77, 293, 104   d 506, 605, 650, 560
   e 845, 8876, 98, 1088   f 2303, 3302, 3032, 2033

> In part **a** the number with the smallest number of 10s is 43 so 43 is the smallest number.
> 61 is last as it has the largest number of 10s.

5 Write down the value of the digit 4 in each of the following numbers.
   a 5047   b 6403   c 3304   d 4056   e 48 976

> Write these numbers in place value diagrams. You can see that the 4 in 5047 is in the tens column, so its value is 40.

6 Write down the value of the digit 6 in each of the following numbers.
   a 3607   b 9056   c 6883   d 62 854

7 Use all three of these cards once to write down
   a the largest number you can make
   b the smallest number you can make.

   | 7 | 9 | 4 |

8 Use all three of these cards once to write down
   a the largest number you can make
   b the smallest number you can make.

   | 4 | 0 | 5 |

> 054 is not an allowed number.

9 a Use the digits 3, 7, 9 and 2 once each to make the smallest possible number.
   b Use the digits 5, 6, 2 and 3 once each to make the largest possible number.
   c Use the digits 7, 0, 5 and 1 once each to make the smallest possible number bigger than five thousand.
   d Use the digits 4, 6, 8 and 5 once each to make the largest possible number smaller than six thousand.

10 Look at the numbers in the cloud:

> two thousand four hundred   2040
>   two thousand and forty-four
>   2440   two thousand two hundred and forty

> Write all the numbers in figures, then look at the thousands, then the hundreds and so on.

  a Which is the largest?   **b** Which is the smallest?

11 Find all the numbers that contain the digits 3, 4, 5 once in each. Write these numbers in order of size with the largest first.

1.2 Adding whole numbers

You should know
- pairs of numbers that add up to 10
- the sum of any two numbers less than or equal to 10
- pairs of numbers that add up to 100.

'Know' means you do not have to work out the answer. Test yourself:
1 Write down the number you need to add to each of these to make 10:
   1, 3, 6, 5, 8
2 Write down the number you need to add to each of these to make 100:
   25, 79, 53, 17, 38, 92, 34, 46, 67, 81, 77

Example 3

Work out $5 + 7 + 3 + 8$.

  23

To add a line of numbers, start at the left-hand side:
  $5 + 7 + 3 + 8$
  $12 + 3$
  $15 + 8 = 23$

Check your answer by starting at the other end.

Example 4

Work out
$$\begin{array}{r} 8 \\ 5 \\ 6 \\ +2 \\ \hline \end{array}$$

  21

To add a column of numbers, start at the bottom and, working in your head, add up the column.

$$\begin{array}{r} 8 \\ 5 \\ 6 \\ 2 \\ \hline 21 \end{array}$$ 21, 13, 8

Check your answer by starting at the top and going down.

Remember that the order in which numbers are added does not matter. It can be easier to look for pairs of numbers that add up to 10.

$$2 + 6 + 7 + 8 + 3 = 10 + 10 + 6$$
$$= 26$$

Exam practice 1B

1 Find the value of:

a 1 + 2 + 3 + 4   b 3 + 6 + 5 + 7
c 6 + 8 + 3 + 9   d 6 + 5 + 4 + 3
e 8 + 4 + 2 + 6   f 9 + 4 + 1 + 6
g 5 + 7 + 3 + 6   h 4 + 2 + 3 + 7
i 8 + 6 + 7 + 9   j 8 + 9 + 7 + 5 + 9
k 3 + 4 + 5 + 6 + 7   l 6 + 7 + 4 + 3 + 8

2 Find the value of:

a	b	c	d	e	f
4	7	8	5	6	2
5	6	3	6	8	7
3	6	2	7	7	6
7	5	4	4	5	4
+2	+ 9	+ 7	+ 3	+ 6	+ 5

3 Look at this sequence of numbers:

   6, 13, 20, 27, ...

You get the next number by adding 7 to the number before.
Write down the next four numbers.

4 Look at this sequence of numbers: 6, 14, 22, 30, ...
You get the next number by adding 8 to the number before.
Write down the next four numbers.

5 Alison drew a plan of her garden.
All her measurements are in metres.
Find the distance round the edge of
Alison's garden.

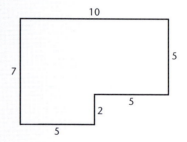

The distance round the edge of a shape is called the **perimeter**.

1.3 Adding larger numbers

It is often easier to add two or more numbers when they are written in a column.

Example 5

Work out 45 + 371 + 647.

Write the numbers in a column.

```
   45
  371
+ 647
-----
 1063
```

First add the units: 13. Write 3 in the units column and carry 1 to the tens column.

Next add the tens: 16. Write 6 in the tens column and carry 1 hundred to the hundreds column.

Then add the hundreds: 10. Write this as 0 hundreds and 1 thousand.

Exam practice 1C

1   Do the following calculations in your head.

a	26	b	25	c	21	d	201	e	53
	+ 53		+ 72		+ 44		+ 107		+ 506

> To find 28 + 51, you can add 1 to 28, then add 50: 28 + 51 = 29 + 50 = 79.

2   Look at these numbers:

19, 23, 31, 41, 62, 69, 89

a   Which two have a total of 50?

b   Which two have a total of 100?

3   Look at these numbers:

31, 41, 47, 62, 53, 79, 89

a   Which two have a total of 100?

b   Which two have a total of 120?

A01 4   Gemma wrote 143 + 426 = 567
Is Gemma correct?
Give a reason for your answer.

> Give a reason for your answer means write down why you think Gemma is correct or not correct.

5   Work out:

a	112	b	113	c	381	d	354	e	441
	14		415		521		17		2046
	+ 273		+ 251		+ 305		+ 614		+ 815

> You do not have to do these in your head, but you must not use a calculator.

f	77	g	8115	h	2011	i	82	j	5844
	103		924		94		189		279
	66		+ 4218		+ 6041		+ 1313		+ 6417
	+ 303								

6   Find:

a   27 + 72 + 13   b   46 + 20 + 92

c   34 + 73 + 76   d   72 + 78 + 27

e   356 + 131 + 302   f   116 + 430 + 452

g   214 + 243 + 394   h   833 + 132 + 159

i   3406 + 367 + 6906   j   2076 + 346 + 569

> Write these numbers in columns like the questions above. Make sure that the tens and units are lined up.

7   There are 301 students in Year 7, 273 in Year 8 and 269 in Year 9. How many students are there altogether in Years 7, 8 and 9?

8   When Toby went to school this morning he walked for 3 minutes to the station. He waited 11 minutes for the train and the train journey took 28 minutes. He then had a 5 minute walk to his school.
How long did it take Toby to get to school?

9   Find the sum of one thousand and fifty, four hundred and seven, and three thousand five hundred.

> Write the numbers in figures.

10

Darts landing in this ring score **double**

Darts landing in this ring score **nothing**

Darts landing in these rings score **single**

Darts landing in this ring score **treble**

Find the score on each board.

a   b   c

d   e   f

1.4 Subtraction

Subtraction and addition are closely related.

If you take 12 away from 75 the answer is 63.
If you then add 12 to 63 the answer is 75.
You can see this on a **number line**.

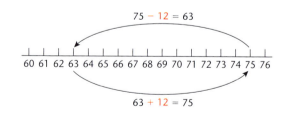

$75 - 12 = 63$

60 61 62 63 64 65 66 67 68 69 70 71 72 73 74 75 76

$63 + 12 = 75$

Example 6

Find $413 - 251$.

$$
\begin{array}{r}
413 \\
- \ 251 \\
\hline
162 \\
\end{array}
$$

You can use any method that you are used to.

You can check your answer by addition:
$162 + 251 = 413$ ✓

Remember that the order in which you add numbers does not matter:
6 + 13 is the same as 13 + 6.

But when you are subtracting the order is important:
7 − 5 is not the same as 5 − 7.

Exam practice 1D

1 Write down the value of:

a	16	b	18	c	19	d	13	e	16	f	20
	− 5		− 6		− 5		− 8		− 9		− 7

2 Work out:

   a 17 − 2   b 11 − 8   c 14 − 9
   d 16 − 7   e 18 − 7   f 14 − 8
   g 726 − 345   h 677 − 341   i 444 − 392
   j 1737 − 635   k 424 − 354   l 603 − 599
   m 895 − 659   n 4003 − 638   p 5400 − 678

3 There are 792 students in a school. 491 are girls. How many boys are there?

4 Take three hundred and forty-one away from four hundred and thirty.

A01

5 a Jane wrote

       354 - 216 = 142

   Without working out the subtraction explain why Jane is wrong.

   b Pete wrote

       246 - 77 = 199

   Is Pete correct? Give a reason for your answer.

> You can check Jane's answer using addition, or work out the units digit.

6 One shop stocks 119 different birthday cards and another shop stocks 155 different cards.
   Find the difference in the number of different birthday cards stocked by the two shops.

> The **difference** between two numbers is the answer when the smaller number is subtracted from the larger one.

7 Subtract three thousand and sixty-four from nine thousand, five hundred and forty-seven.

> First write the numbers in figures, then subtract.

8 What is the difference in the value of the 6 in the number 362 and the value of the 3 in the number 324?

9 The sequence of numbers 100, 92, 84, . . . is formed by subtracting 8 each time.
   Write down the next five numbers in this sequence.

10 Peter needs to score 301 to win a game of darts.
    He scores 55 on his first turn and 87 on his second.
    What does he still have to score to win?

11 Complete these magic squares using the numbers 1 to 9 once only in each square.

a

2	7	6
4		8

b

4		
	5	
	1	6

> In a magic square the total in each row, column and diagonal must be the same.

12 Find the missing digit in each calculation.

a $23 + 52 = \square 5$

b $26 + 6\square = 89$

c $4\square + 36 = 82$

d $\square 6 + 39 = 75$

e $254 + \square 68 = 1022$

f $137 - \square 9 = 78$

g $7\square - 19 = 53$

h $1\square 6 + 284 = 470$

1.5 Mixed calculations

When a calculation has addition and subtraction, e.g. $12 - 15 + 6$, you can change the order and do the addition first.

Example 7

Work out $12 - 15 + 6$.

$$12 - 15 + 6 = 12 + 6 - 15$$
$$= 18 - 15$$
$$= 3$$

> You can change the order to make the calculation easier. Remember that each + or − sign applies to the number **after** it.

Exam practice 1E

1 Find:

a $15 - 6 + 7 - 9$

b $26 - 12 + 3 - 9$

c $7 - 4 + 5 - 6$

d $2 + 13 - 7 + 3 - 8$

e $9 - 12 + 8 - 3$

f $27 + 6 - 11 - 9$

g $17 - 9 + 11 - 19$

h $51 - 27 - 38 + 14$

> When you have a mixture of additions and subtractions you can do all the additions first. Each + or − sign applies to the number **after** it.

2 Jon had 52 DVDs. He gave 23 away and bought another 7. How many DVDs does Jon have now?

 3 Sid wrote $15 - 3 + 7 = 5$.
Sid's answer is wrong. Explain his mistake.

 4 Emma wrote $23 - 8 + 7 = 8$.
Emma's answer is wrong. Explain her mistake.

5 Work out:
   a $200 - 37 + 98 - 31$
   b $300 - 155 + 75 - 140$
   c $61 - 38 + 219 - 84$
   d $29 - 93 + 108 - 9$
   e $95 - 161 + 75 + 10$
   f $95 - 101 - 251 + 438$

6 Four possible answers are given for each calculation. Three of them are obviously wrong and one is correct. Without working them out, write down the letter of the correct answer.
   a $252 - 45$:   **A** 27   **B** 213   **C** 207   **D** 117
   b $684 + 397$:   **A** 287   **B** 1081   **C** 313   **D** 981

1.6 Word problems

Word problems do not tell you if you have to add, subtract, multiply or divide. The clues are in the words. You must read the problem carefully and write your own calculation based on the information given.

Class discussion

Will you need to use addition, subtraction or both to solve these problems? What are the clues that tell you what to do?

1 Mandy had 493 copies of *Maths is Easy* in stock.
  She bought another 500 copies. She sold 650 copies.
  How many does she now have in stock?

2 At the end of last season a football supporters' club had 3459 members.
  This year 774 members joined and 953 members left.
  How many members do they have now?

3 The counter on Gina's car showed a total mileage of 32 743 at the end of Wednesday.
  Gina did 143 miles on Monday, 55 miles on Tuesday and 83 miles on Wednesday.
  What was the reading when she set out on Monday morning?

4 I have a piece of string 200 cm long. I cut off two pieces. One piece is 86 cm long and the other is 34 cm long.
  How long is the piece of string that I have left?

Exam practice 1F

1 1000 portions of curry were cooked in the school kitchen on Monday.
   384 portions were served at the first sitting.
   298 portions were served at the second sitting.
   How many portions were left?

2 Simon has 28 kg of apples when he opens his market stall.
   During the day he gets a delivery of 20 kg of apples and sells 36 kg.
   How many kilograms of apples are left when he closes his stall?

3 An evening class started the year with 42 students.
   During the year 16 students left and 18 joined.
   How many students were there at the end of the year?

4 In a local election one candidate got 526 votes, and the other candidate got 735 votes.
20 voting papers were spoiled and 250 people did not vote.
How many people altogether could have voted?

5 The sequence of numbers 7, 16, 11, 20, 15, … is made by starting with 7, adding 9 to get the next number, subtracting 5 to get the next number, adding 9, subtracting 5, and so on.
a Write down the next 4 numbers in the sequence.
b Starting with 7, write down the third, fifth and seventh numbers.

6 Each part of this ladder is 200 cm long.
There is an overlap of 20 cm at each junction. How long is the extended ladder?

7 In a mathematics book the answers start on page 126 and end on page 134.
How many pages of answers are there?

> Take care! The answer is not 134 − 126.

8 A cedar tree was planted in the year in which Lord Toff was born.
He died on 31st December 1980, aged 90.
a In what year was Lord Toff born?
b How old was the tree in 2005?

9 There are 800 pupils in the village school.
There are 60 more girls than boys. How many of each are there?

Mini coursework task

Copy and complete this magic square using the numbers 1 to 9 once only in each square.

7		3

> The total in each row, column and diagonal must be the same.

1.7 Multiplying by numbers less than ten

Multiplication is the same as adding the same number several times, 5 × 6 means '5 lots of six', or 6 + 6 + 6 + 6 + 6.

You need to know the multiplication tables to multiply accurately without using a calculator.

Test yourself: write down the values of
7 × 4, 6 × 9, 5 × 9, 8 × 3, 9 × 2, 4 × 6, 3 × 5, 2 × 7.

Class discussion

Some of the multiplication facts are harder to remember than others.
Discuss ways of finding:
a 8 × 7   b 6 × 7
c 6 × 8   d 14 × 9
e 17 × 8   f 36 × 6
g 28 × 4

Exam practice 1G

1   For a party, Stella bought 8 of these packets.
    How many balloons did she get?

2   David has to write 80 words on the subject of pets.
    He writes 9 words per line and has written 7 lines.
    a   How many words has he written?
    b   How many more words does he need?

3   Rashida swims 8 lengths each week.
    How many lengths does she swim in 5 weeks?

4   Find:
    a 76×4 b 83×5 c 211×9
    d 204×8 e 142×6 f 305×7

> You can find 76×4 in your head.
> First find 70×4, then 6×4.
> Now add the answers.

5   One jar of jam weighs 516 grams.
    What is the weight of 7 jars of jam?

6   The sequence of numbers 3, 12, 48, … is formed by multiplying
    the last number by 4.
    Write down the next two numbers in the sequence.

7   A sequence of numbers is formed by starting with 7 and
    multiplying by 2 each time.
    Write down the first five numbers in the sequence.

8   Write down the missing digit in these multiplications:
    a $3\boxed{} \times 5 = 185$ b $17 \times \boxed{} = 51$ c $\boxed{}3 \times 8 = 184$

1.8 Multiplication by 10, 100, 1000, …

When a number is multiplied by 10, the digits
move 1 column to the left on a place value
diagram.
The numbers get bigger.

hundreds	tens	units
	8	5
8	5	0 —— 85×10

When a number is multiplied by 100
the digits move 2 columns to the left
on a place value diagram.
The numbers get bigger.

thousands	hundreds	tens	units
		8	5
8	5	0	0 —— 85×100

When a number is multiplied by 1000
the digits move 3 columns to the left
on a place value diagram.
So $85 \times 1000 = 85\,000$.

Example 8

Find 85×20.

$85 \times 2 = 170$

$170 \times 10 = 1700$

So $85 \times 20 = 1700$.

> You can multiply by 20 in two stages:
> multiply by 2 then multiply the answer by 10.

Example 9

Calculate 27×6000.

$27 \times 6 = 162$

$162 \times 1000 = 162\,000$

So $27 \times 6000 = 162\,000$.

> To multiply by 6000, multiply by 6 then multiply the answer by 1000.

 Exam practice 1H

1 Find:
  a 45×10 **b** 63×100 **c** 17×10
  d 26×1000 **e** 56×100 **f** 150×100
  g $37 \times 10\,000$ **h** 250×100 **i** 84×1000

2 Work out:
  a 23×20 **b** 32×200 **c** 43×30
  d 57×50 **e** 24×400 **f** 81×60
  g 57×800 **h** 72×900 **i** 690×600
  j 78×5000 **k** 300×165 **l** 400×499

> 300×165 is the same as 165×300.

3 A box holds 64 bottles.
   How many bottles are there in 500 boxes?

4 Two of these answers are obviously wrong. Which are they?
   $253 \times 2000 = $ **A** $50\,600$ **B** $506\,000$ **C** $5\,060\,000$
   Give a reason for your answer.

 5 There are 150 nails in each packet.
   Pip needed to know the number of nails in 600 packets.
   Which of these answers are obviously wrong?
   A $900\,000$ **B** $75\,000$ **C** $90\,000$
   Give a reason for your answer.

6 A manufacturer supplies reels of cotton in boxes.
   Each box contains 100 reels.
   How many reels will there be in 80 boxes?

 7 Jill was asked how many balloons there were in
   400 of these packets.
   She said,
   '5 times 4 is twenty, then add two noughts for 400.
   There are 2000 balloons in total.'
   What is wrong with her reasoning?

1.9 Long multiplication

Long multiplication is a method for multiplying larger numbers together.

Example 10

Calculate 84 × 26.

```
       84
     × 26
   ──────
    1680   ← 84 × 20
   + 504   ← 84 × 6
   ──────
    2184
```

> Long multiplication method breaks a multiplication down into stages: multiply by the tens, then the units. Lastly add up your answers.

Exam practice 1l

1 Work out:

   a 32 × 21   b 86 × 15   c 34 × 42   d 107 × 26
   e 53 × 62   f 43 × 13   g 27 × 21   h 38 × 41
   i 74 × 106   j 251 × 28   k 46 × 34   l 305 × 19

2 Here is a list of numbers:

   16   23   24   28   33   44   53

   What is the largest number you can make when you multiply two of these numbers together?

3 There are 23 rows of chairs in a school hall with 31 chairs in each row.
   How many chairs are there altogether?

4 A supermarket takes delivery of 54 crates.
   Each crate holds 48 cans.
   How many cans are delivered?

5 Nia counted 33 bags of oranges in a shop.
   There were 16 oranges in each bag.
   How many oranges were there altogether?

6 James buys 12 packs of screws.
   Each pack holds 25 screws.
   How many screws did James buy?

7 Thirteen teams entered a tug-of-war competition.
   Each team had 21 players.
   How many players were there in total?

1.10 Division by a whole number less than ten

$32 \div 8$ means 'How many eights are there in 32?'
You could find out by repeatedly taking 8 away.

$32 - 8 = 24$
$\qquad 24 - 8 = 16$
$\qquad\qquad 16 - 8 = 8$
$\qquad\qquad\qquad 8 - 8 = 0$

so there are 4 eights in 32.

It is quicker if you know the multiplication facts, because
multiplication and division are the opposite of each other:

$32 = 4 \times 8$, so $32 \div 8 = 4$.

You can use this fact to check your answers.

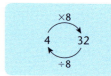

Example 11

Calculate $534 \div 3$.

$534 \div 3 = 178$

Check: $178 \times 3 = 534$ ✓

Start with the hundreds:
5 (hundreds) ÷ 3 = 1 (hundred), remainder 2 (hundreds)

Add this remainder to the tens:
23 (tens) ÷ 3 = 7 (tens), remainder 2 (tens)

Add this remainder to the units:
24 (units) ÷ 3 = 8 units

Exam practice 1J

1   Work out:
   a $58 \div 2$ b $84 \div 4$ c $618 \div 6$ d $1764 \div 6$
   e $1773 \div 9$ f $366 \div 6$ g $605 \div 5$ h $352 \div 8$
   i $7355 \div 5$ j $2478 \div 7$ k $2118 \div 6$ l $508 \div 4$

2   A greengrocer bought a sack of potatoes weighing 50 kg.
    He divided the potatoes into bags, so that each bag held 2 kg of
    potatoes.
    How many bags of potatoes did he get from his sack?

3   There are 192 seats in a hall.
    They are arranged in rows of 8 seats.
    How many rows are there?

4   A lorry is loaded with boxes.
    The total weight of the boxes is 702 kg.
    Each box weighs 9 kg. How many boxes are there?

 5   A college wants to take 376 students to the theatre in buses.
    Each bus takes 47 students.
    Penny says that they will need 8 buses.
    Is she right? Give a reason for your answer.

6   How many fives can be taken away from 135?

Class discussion

How do you know the
value of
$63 \div 9$, $49 \div 7$,
$56 \div 8$, $27 \div 3$,
$36 \div 9$, $24 \div 8$,
$28 \div 7$, $54 \div 6$,
$42 \div 6$, $18 \div 2$,
$30 \div 6$, $72 \div 8$?

1.11 Remainders

When you divide a whole number by another whole number, the answer may not be exactly a whole number.
For example, 12 ÷ 5 means 'How many fives are there in twelve?'
You know that there are 2 fives in twelve with 2 units left over.
These two units are called the remainder.
This is written as 12 ÷ 5 = 2, remainder 2.

Example 12

June has 50 loose eggs and some egg boxes.
One egg box holds 12 eggs.
a   How many egg boxes can June fill?
b   How many egg boxes does June need to pack all the eggs?

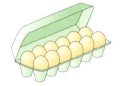

a   50 ÷ 12 = 4, remainder 2.
    June can fill 4 boxes.

> There are 4 complete twelves in 50, so 4 boxes can be filled. The two eggs left over will not fill a box.

b   She needs 5 egg boxes.

> To put all the eggs into boxes, June needs another box for the 2 eggs left over.

Exam practice 1K

1   Work out:
    a   75 ÷ 4   b   85 ÷ 7   c   731 ÷ 6
    d   715 ÷ 8   e   913 ÷ 5   f   565 ÷ 9
    g   133 ÷ 6   h   442 ÷ 4   i   534 ÷ 7
    j   386 ÷ 5   k   333 ÷ 5   l   783 ÷ 7

> Remember to give the remainder.

2   Ben has 166 apples. He puts them in packs of 6.
    a   How many packs can he make up?
    b   How many apples are left over?

3   A factory has 378 car wheels in stock.
    Each car needs 5 wheels.
    a   How many cars can they fit from stock?
    b   How many wheels will be left over?

4   Molly has 405 leaflets to deliver.
    She must put 6 into each house.
    a   How many houses can she deliver to?
    b   How many leaflets are left over?

5   At an athletics meeting, 84 people enter for a race.
    Several rounds must be run, with the same number of runners in each round.
    There are ten lanes. How many rounds will be needed?

6 Sara has 95 eggs.
  She puts them into boxes. Each box holds 6 eggs.
  a How many boxes can Sara fill?
  b She puts the eggs that are left over in an empty box.
     How many more eggs can she put in this box?

7 5 sacks of potatoes are delivered to a shop.
  Each sack holds 50 kg.
  The potatoes are put in bags with 4 kg in each bag.
  a Work out the total weight of the delivery.
  b How many 4 kg bags can the shop make up?

8 Sixty-six people travel by taxi.
  Five people can fit into each taxi.
  a How many taxis are needed?
  b How many spare seats will there be?

9 Mr Brown has 437 books.
  He gives 3 books to every student in his class.
  There are 2 left over.
  Work out the number of students in his class.

10 A school has 859 exercise books to give out to Year 11.
  This is enough for every pupil to have 8 exercise books.
  Fewer than 8 exercise books are left over.
  How many students are there in the year?

1.12 Division by 10, 100, 1000, …

Division is the opposite of multiplication.

When you divide by 10 the digits move one column to the right on a place value diagram. The numbers get smaller.

When you divide by 100 the digits move two columns to the right on a place value diagram. The numbers get smaller.

When you divide by 1000 the digits move three columns to the right on a place value diagram. The numbers get smaller.

Any digits to the right of the units column are the remainder.

hundreds	tens	units		
8	1	2		
		8	1	2

$\leftarrow 812 \div 100 = 8$ remainder 12

Exam practice 1L

1 Work out:
  a $256 \div 10$ **b** $87 \div 10$ **c** $942 \div 100$
  d $3077 \div 100$ **e** $876 \div 100$ **f** $4910 \div 1000$
  g $2781 \div 10$ **h** $8512 \div 100$ **i** $7230 \div 1000$

2 3600 exercise books are wrapped into packs with 100 books in each pack. How many packs are made?

3 Twenty-four thousand beads are put into bags.
A hundred beads can fit in each bag.
How many bags are needed?

4 2440 pens are packed into boxes.
There are 100 pens in each box.
  a How many full boxes are there?
  b How many pens are left over?

5 1000 sweets are packed into bags.
Each bag contains 40 sweets.
How many bags are needed?

> You can divide 1000 by 10, then divide the answer by 4.

6 750 students are put into football squads.
There are 30 students in each squad.
How many squads are there?

7 Sixty thousand pens are put into packets.
Each packet takes 20 pens.
How many packets can be filled?

8 A £2 000 000 lottery jackpot is divided equally between 40 winners.
How much does each winner get?

9 Six hundred and sixty people get on a train.
Everyone gets a seat but there are no empty seats.
Each carriage has 60 seats.
How many carriages are there?

1.13 Long division

You can use the same method for long division as you use for short division.

Example 13

Find $2678 \div 21$.

```
        1 27
   21) 2678
       21
       57
       42
      158
      147
       11
```

Start by dividing 21 into 26 hundreds.

This is equal to 1 twenty-one, remainder 5 (hundreds).

Add this remainder to the tens

Divide 21 into 57. This is 2 twenty-ones (42), remainder 15 (tens).

Add this remainder to the units.

Divide 21 into 158. This is 7 twenty-ones (147), remainder 11 (units).

This is the remainder.

$2678 \div 21 = 127$, remainder 11.

Exam practice 1M

1   Work out:
   a   254 ÷ 20   b   875 ÷ 25   c   7514 ÷ 24   d   8013 ÷ 40
   e   685 ÷ 13   f   269 ÷ 16   g   6372 ÷ 27   h   829 ÷ 16

> Remember to give the remainder.

2   264 books are stacked on shelves in piles of 12.
    How many piles are there?

3   500 supporters travel to a match in coaches.
    Each coach carries 47 supporters.
    a   How many coaches are needed?
    b   How many empty seats will there be?

4   4000 apples are packed into boxes.
    Each box holds 75 apples.
    How many boxes are needed?

5   Grapefruit are packed 15 to a box.
    a   How many boxes are needed to pack 250 grapefruit?
    b   How many more grapefruit are needed to fill the last box?

6   Apples are packed 14 to a bag.
    a   How many bags are needed to pack 284 apples?
    b   How many more apples are needed to fill the last bag?

7   Tea bags are packed 48 to a box.
    2055 tea bags are to be boxed.
    a   How many boxes are needed?
    b   How many more tea bags are needed to fill the last box?

Mini coursework task

Jo has some marbles and some bags to keep them in.
If 8 marbles are put into each bag, 4 marbles are left over.
If 10 marbles are put into each bag, one bag is empty.
How many marbles and how many bags are there?
Explain your answer.

> Make sure that what you write makes sense. Also make sure that anyone else who reads your work can understand it.
> You can start by discussing with others how to tackle this problem.

1.14 Mixed operations

When a calculation is a mixture of addition and subtraction, and multiplication and division, you always do the multiplication and division first.

Example 14

Find $2 \times 4 + 3 \times 6$.

$$2 \times 4 \; + \; 3 \times 6 = 8 + 18$$
$$= 26$$

> Find 2×4 and 3×6 first. Then add the answers.

Example 15

Find $8 - 6 \div 2$.

$8 - 6 \div 2 = 8 - 3$
        $= 5$

Find $6 \div 2$ first.

Exam practice 1N

1   Find:
  a $2 \times 4 + 8$ **b** $3 \times 6 - 2$
  c $10 - 2 \times 3$ **d** $15 - 2 \times 7$
  e $2 + 4 \times 6 + 8$ **f** $24 \div 8 - 2$
  g $7 - 3 \times 2 + 8 \times 2$ **h** $7 + 4 - 3 \times 2$

2   Work out:
  a $7 \times 2 + 6 - 1$ **b** $8 + 3 \times 2 - 4 \div 2$
  c $6 \div 2 + 6 \times 3$ **d** $45 \div 5 - 3 \times 3$
  e $6 + 8 \div 4 + 6 \div 2$ **f** $14 \times 2 \div 7 - 3$
  g $5 \times 3 \times 4 \div 12$ **h** $9 \div 3 + 6 \times 2$
  i $8 + 16 \div 4 - 3$ **j** $18 \div 6 + 18 - 6$

1.15 Brackets

Brackets are used to show working out that needs to be done first.
For example $(11 + 7) \times 10$ means add 11 and 7 first, then multiply
the answer by 10.

Example 16

Find   **a** $8 \times (5 - 3)$ **b** $(14 - 2) \div 3$.

a $8 \times (5 - 3) = 8 \times 2$
              $= 16$

Work out $5 - 3$ first.

b $(14 - 2) \div 3 = 12 \div 3$
                $= 4$

Work out $14 - 2$ first.

Exam practice 1P

1   Find:
  a $4 \times (2 + 3)$ **b** $(7 + 3) \times 18$ **c** $(8 - 5) \times 3$
  d $(2 - 1) \times 6$ **e** $(3 + 9) \times 10$ **f** $(6 - 2) \div 2$

2   Work out:
  a $5 \times (12 - 7)$ **b** $4 \times (7 + 13)$ **c** $(12 - 7) \div 5$
  d $(1 + 7) \div 4$ **e** $8 \times (1 + 4)$ **f** $(46 - 16) \div 6$

3   Find **a** $(2 + 8) \times 4$ **b** $2 + 8 \times 4$

Summary of key points

- The position of each digit in a number tells you its value
  e.g. in 5273 the 7 is 7 tens and the 5 is 5 thousands.
- It is easier to add or subtract larger numbers when they are written in columns.
- The order in which you add or multiply numbers does not matter.
- The order in which you subtract or divide numbers is important.
- In questions with mixed operations you must do multiplication and division before addition and subtraction.
- Read questions carefully for the clues that tell you whether to add, subtract, multiply or divide.

Most students who get GRADE E or above can:
- deal with problems involving remainders.

Most students who get GRADE C can also:
- work with a mixture of addition, subtraction, multiplication and division.

Glossary

Difference	the answer when the smaller of two numbers is taken away from the larger one
Digit	any of the single figures from 0 to 9, e.g. 2, 5, 6
Number line	a line like this, marked with numbers:

60 61 62 63 64 65 66 67 68 69 70 71 72 73 74 75 76

Perimeter	the distance round the outside of a shape
Place value	the value of each figure in a number

2 Negative numbers

2.1 Negative numbers

Negative numbers are written with a minus sign in front, for example, you write negative 2 as −2.

Negative numbers can be used to describe the distance below sea level or temperatures lower than 0°C.

Numbers greater than a natural zero are called **positive numbers**.

> 2 is a positive number. You can write 2 as +2, but the + is usually left out.

Positive and negative numbers together are known as **directed numbers**.

A line marked with positive and negative numbers is called a **number line**.

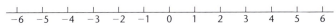

On this number line, 4 is to the *right* of 2

> > means 'is greater than'.

so   4 > 2

                2 is to the *left* of 6

> < means 'is less than'.

so   2 < 6

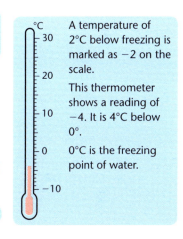

°C

A temperature of 2°C below freezing is marked as −2 on the scale.

This thermometer shows a reading of −4. It is 4°C below 0°.

0°C is the freezing point of water.

Example 1

Which is larger   **a** 1 or −2   **b** −2 or −4?

a 1

> 1 is larger than −2 because 1 is to the right of −2 on the number line.

b −2

> −2 is larger than −4 because −2 is to the right of −4 on the number line.

Example 2

Write down   **a** the smaller of the two numbers −3 and 2

             b the larger of the two numbers −4 and −1.

a −3

> −3 is to the left of 2 on the number line.

b −1

> −1 is to the right of −4 on the number line.

Exam practice 2A

1   Write down the larger of each pair of numbers.
    a   3, −2   b   −1, 0   c   −3, −1

2   Write down the smaller of each pair of numbers.
    a   −3, −9   b   −4, 3   c   2, −2

3   The temperatures in six towns one morning were:
    −7°C   −11°C   −9°C   5°C   7°C   −3°C
    a   Write down the temperature of the coldest town.
    b   Write down the temperature of the warmest town.

4   The temperatures in five villages one day were:
    −8°C   4°C   −9°C   −6°C   −10°C
    a   Write down the temperature of the coldest village.
    b   Write down the temperature of the warmest village.

5   a   The temperatures in eight Swiss towns one morning were:
        −5°C   −10°C   −8°C   5°C   9°C   −3°C   −11°C   3°C
        Write these temperatures in order with the lowest first.
    b   The temperatures each morning one week in Speightown were:
        −7°C   −10°C   −4°C   3°C   −3°C   −8°C   7°C
        Write these temperatures in order with the lowest first.

6   a   The temperatures in seven French towns one day were:
        −12°C   −6°C   4°C   5°C   −9°C   −7°C   −10°C
        Write these temperatures in order with the highest first.
    b   The midday temperatures during one week in Moscow each day were:
        −11°C   −5°C   6°C   5°C   −8°C   −4°C   −13°C
        Write these temperatures in order with the highest first.

7   The map shows the temperatures in some cities in January.

    a   Write down the warmest city.
    b   Write down the coldest city.

> **Did you know**
> that the ancient Chinese would not accept a negative number as a solution to a question or problem? They would change the problem so the answer was positive.

> Copy the list. Now find the lowest temperature, write it down and then cross it off the list. Do this for the next temperature. Carry on like this until all the temperatures have been crossed off the list.

8 a Write these numbers in order with the smallest first.

$-4, -7, 3, 15, -23, -6$

   b Write these numbers in order with the largest first.

$-8, 12, -5, -13, -10, 9$

9 This is a thermometer for a fridge-freezer.

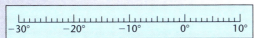

   a The temperature inside the fridge should be 5°C.
      Copy the scale and mark this temperature with an arrow.
   b The temperature in the freezer should be −28°C.
      Mark this temperature with an arrow on your scale.

2.2 Addition and subtraction

When you take away a positive number you can end up with a
negative answer.

Example 3

Use a number line to find $-2 + 5$.

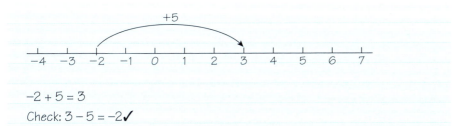

Move 5 places to the right on the number line.

$-2 + 5 = 3$

Check: $3 - 5 = -2$ ✓

Example 4

At midnight, the temperature was 5°C. In the next two
hours the temperature fell by 7 degrees. What was the
temperature at 2 a.m?

You can use a number line.
Start at 5°C and move left 7 places.

−2°C

Check: $-2 + 7 = 5$ ✓

These are the rules for adding and subtracting negative numbers.

**Adding a negative number is the same as
subtracting a positive number.**

**Subtracting a negative number is the same as
adding a positive number.**

Example 5

Find   **a** $1 + (-4)$ **b** $2 - (-5)$.

a $1 + (-4) = 1 - 4$
$\qquad = -3$

Adding -4 is the same as subtracting 4.

b $2 - (-5) = 2 + 5$
$\qquad = 7$

Subtracting -5 is the same as adding 5.

Putting brackets round the negative number make it easier to see the two signs.

Exam practice 2B

1 Find: **a** $3 - 7$ **b** $-5 - 2$ **c** $-4 - 6$
          d $-7 + 3$ **e** $4 - 5$ **f** $-2 + 6$

You can use a number line to help you.

2 Find: **a** $6 + (-5)$ **b** $4 + (-8)$ **c** $8 + (-2)$
          d $12 + (-7)$ **e** $3 + (-6)$ **f** $5 + (-10)$

Adding -5 is the same as subtracting 5.

3 Find: **a** $6 - (-3)$ **b** $-4 - (-2)$ **c** $8 - (-5)$
          d $12 + (-8)$ **e** $-12 - (-8)$ **f** $-12 - 8$

Taking away -3 is the same as adding 3.

4 Find the missing number:
   a $7 - 9 = \square$ **b** $4 - 5 = \square$
   c $3 - 5 = \square$ **d** $-8 - 7 = \square$

5 Find: **a** $7 - 9 + 4$ **b** $10 - 4 - 9$ **c** $-2 - 3 + 9$
          d $-3 - 4 + 2$ **e** $-4 + 2 + 5$ **f** $-5 + 6 - 7$

6 Work out: $-6 + 7 - 8 + 12$

7 When Tom went to bed the temperature was $-6°C$.
By the time he got up it had fallen 4 degrees.
What was the temperature when Tom got up?

8 **a** The temperature was $-6°C$. It went up 7 degrees.
      What is the temperature now?
    b The temperature was $-3°C$. It went down 4 degrees.
      What is the temperature now?

The difference between two numbers is the answer when the smaller number is subtracted from the larger one.

9 Find the difference between:

   a -8 and -3 **b** -4 and 7.

10 When Keri boarded the plane in the Caribbean to come home
the temperature was $26°C$. During the flight the temperature
outside the aircraft was $-57°C$. Find the difference between
these two temperatures.

11 a Look at this sequence of numbers.

$$-10, -7, -4, \ldots$$

To get the next number you add 3 to the number before.
Write down the next two numbers in the sequence.

b Look at this sequence of numbers.

$$8, 5, 2, \ldots$$

To get the next number you subtract 3 from the number before.
Write down the next three numbers in the sequence.

12 The time in Hong Kong is 7 hours ahead of the time in Rome.
  a What time is it in Rome when it is 8 a.m. in Hong Kong?
  b What time is it in Hong Kong when it is 8 a.m. in Rome?

> This means that Hong Kong time is Rome time plus 7 hours.

13 The time in Florida is UK time minus 5 hours.
What time is it in Florida when the time in the UK is

  a 10 a.m.   **b** 8 p.m.   **c** 2 p.m?

14 The time in Athens is UK time plus 2 hours.
What time is it in Athens when the time in the UK is

  a 7 a.m.   **b** 5 p.m.   **c** midnight?

15 The time in Vancouver is UK time minus 8 hours.
Donna lives in the UK.
She arranges to phone her sister in Vancouver, at 2 p.m.
Vancouver time. What is the time in the UK when she should ring?

16 When Freda crossed the International Date Line on a journey from Tokyo to Los Angeles she had to put her clock back 24 hours. Two hours before she reached the Date Line it was 2200 hrs on 5th May. What was the date and time two hours after she had crossed the line?

Mini coursework task

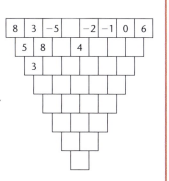

Each number in this triangular pattern is formed by finding the difference between the two numbers directly above it.

Copy the diagram and fill in the empty spaces.
Is there more than one way in which it can be completed?

2.3 Multiplication and division of directed numbers

When a positive number and a negative number are multiplied or divided, the answer is negative.

Example 6

Find: **a** $2 \times (-3)$ **b** $(-8) \div 2$

   a $2 \times (-3) = -6$

> $2 \times (-3)$ means '2 lots of -3'.

   b $(-8) \div 2 = -4$

When two numbers are multiplied or divided and their signs are the same, the answer is positive.

Example 7

Find: **a** $-2 \times (-3)$ **b** $-8 \div (-4)$

   a $-2 \times (-3) = 6$

> Both numbers are negative, so the answer is positive.

   b $-8 \div (-4) = 2$

When two numbers are multiplied or divided:
if the signs are the same, the answer is positive,
if the signs are different, the answer is negative.

Exam practice 2C

1   Find: **a** $3 \times (-2)$ **b** $(-5) \times (-4)$ **c** $(-4) \times (-5)$
               d $(-3) \times (-4)$ **e** $5 \times (-3)$ **f** $(-3) \times 7$
               g $(-7) \times 6$ **h** $5 \times (-10)$ **i** $6 \times (-4)$

2   Find: **a** $8 \div (-2)$ **b** $15 \div (-5)$ **c** $(-12) \div 2$
               d $(-24) \div 6$ **e** $12 \div (-3)$ **f** $20 \div (-4)$
               g $(-16) \div 8$ **h** $(-35) \div 7$ **i** $(-30) \div 6$

3   Work out: **a** $4 + 3 \times (-4)$ **b** $(-2) \times 3 + 4 \times (-5)$
                   c $8 + 2 \div (-2)$ **d** $2 - 4 \times (-2)$

> Remember, do multiplication and division first.

4   Work out: **a** $(-4) \times (-7)$ **b** $8 - (-3) \times 5$
                   c $(2 - 10) \div (-4 \times 2)$ **d** $(-1) \times (-2)$
                   e $12 + 3 \times (-6)$ **f** $(5 - 8) \div (2 - 5)$

> Work out the inside of the brackets first.

5 Work out a $(3 - 5) \div (-2)$
            b $2 \times (-2) \times (-3)$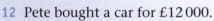
            c $(12 - 24) \div (9 - 3)$

> When you multiply three numbers you can start by multiplying the last two together first.

6 a Find a positive number that when it is multiplied by itself gives:
     i 25   ii 4   iii 16.
  b Find a negative number that when it is multiplied by itself gives
     i 9   ii 16   iii 25.

7 a What number has to be multiplied by -6 to get 42?
  b What number must you divide -20 by to get 5?
  c What number has to be multiplied by -8 to get -32?
  d What number must you divide -18 by to get -6?

8 Look at this sequence of numbers: $1, -3, 9, \ldots$
  To get the next number you multiply the previous number by -3.
  Work out the next three numbers in the sequence.

9 Look at this sequence of numbers: $-4, 8, -16, 32, -64$
  a What do you multiply each number by to get the next one?
  b Write down the next two numbers in the sequence.

10 A gambler started with £1000.
   He lost £300 a night for 5 nights.
   How much money did he have at the end of five nights?

11 Ann entered a quiz where there were 20 questions.
   She scored 3 points for a correct answer but lost a point for each
   wrong answer.
   Ann answered 12 questions correctly and the rest incorrectly.
   How many points did Ann score?

12 Pete bought a car for £12 000.
   Each year for the next 8 years it went down in value by £1400.
   a What was the car worth when it was 8 years old?
   By this time it had become collectable.
   For the next 3 years its value went up by £1000 a year.
   b What was it worth after 11 years?

13 On New Year's Day Ellen had £176 in her bank account.
   Over the next 9 months she put £75 a month into the account.
   a How much did she have in the bank after nine months?
   b For the last three months of the year she took £215 out each
      month from the account.
   How much did Ellen have in the account at the end of the year?

14 In a test students were given 4 marks for a correct answer and -1
   for a wrong answer. Every question had to be attempted.
   a Olive got 18 right and 12 wrong. How many did she score?
   b Zoe got 8 right and 22 wrong. How many did she score?
   c How many questions were there in the test?
   d Edna scored 0. How many did she get right?

Summary of key points

- Adding a negative number is the same as subtracting a positive number.
- Subtracting a negative number is the same as adding a positive number.
- If two numbers are multiplied or divided, then
    if the signs are the same (both + or both −), the result is positive
    if the signs are different (one + and one −), the result is negative.

Most students who get GRADE E or above can:
- understand temperatures below freezing
- add and subtract negative numbers.

Most students who get GRADE C can also:
- multiply and divide directed numbers.

Glossary

Difference	the answer when the smaller of two numbers is subtracted from the larger one
Directed number	positive and negative numbers together are called directed numbers
Negative number	a number below zero, for example negative 3, written as −3
Number line	a line like this marked with positive and negative numbers

Positive number	numbers above zero, for example 2 and 7

3 Types of numbers

This chapter will show you:
✓ how to recognise different types of numbers
✓ tests to find out if one number will divide exactly by another number
✓ what a prime number is
✓ how to find factors of a number
✓ how to find multiples of a number
✓ how to find the highest common factor of two or more numbers
✓ how to find the lowest common multiple of two or more numbers
✓ the meaning of indices and how to use them
✓ the meaning of square root and cube root

Before you start you need to know:
✓ your multiplication tables up to 10×10
✓ how to divide by whole numbers less than 10

3.1 Names for number

The **counting numbers** are 1, 2, 3, 4, 5, 6, 7, ...

> These are also called **natural numbers**.

An **integer** is a whole number. It can be positive or negative and includes 0.

> −9, 413 and 5 are integers.

Integers are either even numbers or odd numbers.

An **even number** ends in 0, 2, 4, 6 or 8.
An even number will divide exactly by 2.

> 16, 58, 656 are even numbers.

An **odd number** ends in 1, 3, 5, 7 or 9.
An odd number will not divide exactly by 2.

> 5, 57, 889 are odd numbers.

A **prime number** can only be divided exactly by 1 and itself.
For example, 5 is a prime number, as 1 and 5 are the only whole numbers that divide into 5 exactly.
1 is *not* a prime number but 2 is a prime number.
2 is the only even prime number.

> When a whole number is divided by a whole number, 'exactly' means the answer is a whole number with no remainder.

Did you know

that prime numbers often occur as consecutive odd numbers e.g. 29 and 31?
There is no pattern to find the next prime number from one you know.
There is no largest prime number.

If any integer is multiplied by itself the answer is a **square number**.

> 25 is a square number as $25 = 5 \times 5$.
> 81 is a square number as $81 = 9 \times 9$.

Exam practice 3A

1 Which of the numbers 17, 20, 46, 57 and 92 are even?

2 Which of the numbers 15, 24, 53, 72 and 193 is
  a the smallest odd number
  b the largest even number?

3 Which of the following numbers are prime numbers?
  a 7, 19, 24, 31, 41
  b 29, 16, 6, 11, 39, 42

4 The even numbers 8 and 16 can be written as the sum of two prime numbers.
8 can be written as 3 + 5 and 16 can be written as 5 + 11.
Write the following even numbers as the sum of two prime numbers.
  a 6   b 10   c 14

5 The odd numbers 13 and 19 can be written as the sum of three prime numbers.
13 can be written as 3 + 5 + 5 and 19 can be written as 3 + 5 + 11.
Write each of the following odd numbers as the sum of three odd prime numbers.
  a 11   b 15   c 17

6 a List all the prime numbers less than 50.

A01

  b The prime numbers 3 and 47 add up to 50.
     Ken said that there are other pairs of prime numbers that add up to 50?
     Is Ken correct? Give a reason for your answer. ●————

> Give a reason for your answer means write down
> 'Yes because...(two prime numbers that you have found) add up to 50.' or
> 'No, because no two of these numbers...(list of the other prime numbers less than 50) add up to 50.'

7 Which of the numbers 4, 8, 16, 30 and 36 are square numbers?

A01

8 Nicky said that 100 is a square number.
Is Nicky correct?
Give a reason for your answer.

A01

9 Chris said that 56 is a square number.
Is Chris correct?
Give a reason for your answer.

10 Look at the numbers in this circle.
  a Two of these numbers are odd numbers. Write them down.
  b Two of these numbers are prime numbers. Write them down.
  c There is one square number. Write it down.

11 For each of these statements, give an example to show that it is not true.
  a All the prime numbers are odd numbers.
  b When an even number is divided by 2 it gives an odd number.

3.2 Factors

A **factor** of a number divides into the number exactly leaving no remainder.

Example 1

Write down all the factors of 12.

$1 \times 12 = 12$

$2 \times 6 = 12$

$3 \times 4 = 12$

The factors of 12 are 1, 2, 3, 4, 6 and 12.

> Look for pairs of numbers that multiply to give 12.

> No other whole numbers will divide into 12 exactly leaving no remainder.

A prime number which is a factor of a number is called a **prime factor**.

Example 2

Lucy said that both 3 and 4 are prime factors of 24.
Lucy is wrong. Explain why Lucy is wrong.

4 is a not a prime factor of 24 because 4 is not a prime number.

The following tests for dividing by different numbers are useful.
- An even number will divide by 2.
- A number whose digits add up to a number that can be divided exactly by 3 can also be divided exactly by 3.
  So 171 can be divided exactly by 3 as $1 + 7 + 1 = 9$ and $9 \div 3 = 3$.
- A number ending in 5 or zero can be divided exactly by 5.
  So 230 and 155 can be divided exactly by 5.
- A number whose digits add up to a number that can be divided exactly by 9 can also be divided exactly by 9.
  So 261 can be divided exactly by 9 as $2 + 6 + 1 = 9$ and $9 \div 9 = 1$.

> If the sum of the digits is not divisible by 3, the number itself is not divisible by 3; a number must end in 0 or 5 to be divisible by 5; if the sum of the digits is not divisible by 9, the number is not divisible by 9.

Exam practice 3B

1 Find all the pairs of numbers that multiply to give:

   a 8   **b** 12   **c** 6   **d** 16   **e** 18   **f** 24

 2 John said that 78 and 48 can be divided by 3 exactly.
Is John correct? Give a reason for your answer.

 3 Will each of the following numbers divide exactly by 3?
Give a reason for your answer.

   a 177   **b** 84   **c** 135   **d** 266   **e** 4056   **f** 219

> For example you can write 20 as 1×20, 4×5 or 2×10. The factors of 20 are 1, 2, 4, 5, 10 and 20.

> If you add up the digits of a number and the total will divide exactly by 3, the number you started with will divide by exactly 3.

4   Colin said that 207 and 576 will both divide exactly by 9.
    Is Colin correct? Give a reason for your answer.

> If you add up the digits of a number and the total will divide exactly by 9, the number you started with will divide exactly by 9.

5   Which of these numbers divides exactly by 9?
    a   36   b   369   c   468   d   725   e   558

6   Find two different numbers that are factors of 15.

> You need any whole number that divides into 15 without leaving a remainder. There are four such numbers.

7   Find three different numbers that are factors of:
    a   24   b   26   c   14

> Remember that 1 is a factor of any number and so is the number itself.

8   a   What are the two prime factors of 18?
    b   Find all the prime factors of: i 21   ii 42.

9   Find all the factors of:
    a   18   b   28   c   38

10 a   Kwame said that the factors of 20 are 1, 2, 4, 5 and 10.
        Is Kwame correct? Give a reason for your answer.
    b   Jim said that the only factors of 10 are 1, 2 and 5.
        Is Jim correct? Give a reason for your answer.

11 Look at this list of numbers: 2, 19, 31, 40, 54, 65
    Write down:
    a   the smallest number exactly divisible by 5
    b   the largest prime number
    c   the smallest number exactly divisible by 3.

12 Which of the following statements are true?
    Give reasons for your answers.
    a   All prime numbers between 10 and 100 are odd numbers.
    b   All odd numbers are prime numbers.
    c   The number 7893 is exactly divisible by 9.

> To show that a statement is false you need to find *one* example that does not work.

3.3 Multiples of a number

When a number is multiplied by a whole number, the answer is a **multiple** of the first number.

You can recognise the multiples of 3 from the 3 times table.
So 3, 6, 9, 12, 15, 18, 21, 24, 27, 30, ... are all multiples of 3.

You can tell if a larger number is a multiple of 3, because 3 will divide exactly into it. 300 and 153 are multiples of 3.

Example 3

Is 30 a multiple of 10? Give a reason for your answer.

Yes because 3 × 10 = 30.

> Another reason is that 30 ÷ 10 = 3 with no remainder.

Exam practice 3C

1   a   Write down a multiple of 7.
    b   Write down two multiples of 9.
    c   Write down three multiples 12.

2   Give a reason for your answer to each of the following questions.
    a   Is 20 a multiple of 5?
    b   Is 54 a multiple of 7?
    c   Is 46 a multiple of 6?
    d   Is 56 a multiple of 8?

3   Look at this list of numbers:
    28, 36, 42, 66, 72
    a   Which number is a multiple of 7?
    b   Which number is a multiple of 9?
    c   Write down the number that is not a multiple of 6.

A01   4   Kim said that 108 is a multiple of 9.
    Is Kim correct?
    Give a reason for your answer.

A01   5   Polly said that 234 is a multiple of 3.
    Is Polly correct?
    Give a reason for your answer.

A01   6   David said that 126 is a multiple of 7.
    Is David correct?
    Give a reason for your answer.

3.4 Common factors

Two or more numbers can have the same factor.
This is called a **common factor**.

Example 4

Find the common factors of 21 and 28.

The factors of 21 are **1**, 3, **7**, 21
The factors of 28 are **1**, 2, 4, **7**, 14, 28
1 and 7 are the common factors of 21 and 28.

> List the factors of each number.
> The common factors are the numbers that are in both lists.

Highest common factor

The **highest common factor** of two or more numbers is the *largest* whole number that divides exactly into all of them.

You can find the HCF by writing down the factors of each number.

> 'Highest common factor' is written as HCF.

Example 5

Find the highest common factor of 56 and 52.

Factors of 56: 1, 2, **4**, 7, 8, 14, 28, 56
Factors of 52: 1, 2, **4**, 13, 26, 52

> List the factors of each number.

4 is the HCF of 56 and 52.

> 4 is the highest number in both lists.

Exam practice 3D

1   Find the common factors of 35 and 56.

2   Find the largest number that is a factor of both 9 and 12.

 3   a   Is 9 a common factor of 27 and 38?
          Give a reason for your answer.
    b   Rose said that 8 is a common factor of 32 and 48.
          Is Rose right? Give a reason for your answer.

4   Write down the largest whole number that will divide exactly into

    a   9 and 12   b   8 and 16   c   12 and 24.

5   Write down the largest whole number that will divide exactly into

    a   25, 50 and 75   b   22, 33 and 44
    c   21, 42 and 84   d   39, 13 and 26
    e   25, 35, 50 and 60   f   36, 44, 52 and 56.

6   a   Find the highest common factor of 72 and 20.
    b   Use your answer to **a** to find the largest whole number that will divide exactly into 720 and 200.

7   Find the largest whole number that will divide exactly into 360 and 480.

 8   George said that 8 is the largest whole number that will divide exactly into 36, 44, 52 and 56.
    Explain why George is wrong.

9   A rectangular floor measures 450 cm by 350 cm.
    The floor needs to be covered with square tiles.
    a   What is the largest tile size that could cover the floor exactly?
    b   How many tiles of this size are needed?

> **Class discussion**
> The HCF of two numbers cannot be bigger than the difference between the two numbers.
> Explain why this is true.
> Can you think of a similar rule for the HCF of three numbers?

> All four sides of a square are the same length.

10 Harry has a rectangular piece of chipboard measuring 42 cm by 30 cm.
    He wants to divide it into identical squares.
    a   What is the largest square size he could use.
    b   How many square tiles will he produce?

11 105 apples and 63 oranges are shared among some students.
    Work out the largest number of students that can get equal shares of both apples and oranges.

12 A rectangular-shaped patio measures 550 cm by 330 cm.
    a   What is the largest size of square stone that can be used to pave the area without any cutting?
    b   How many of these stones will be needed?

13 Hurdles are used to fence off a rectangular field measuring 339 metre by 273 metres.
    The hurdles are all the same length, and must not overlap.
    Find the longest hurdle that can be used.

14 A floor measures 504 cm by 396 cm.
    It is to be covered with square tiles.
    Only whole ones are to be used.
    a   What is the greatest size they can be?
    b   How many tiles will be used?

3.5 Common multiples

One number can be a multiple of two or more numbers.
This is called a **common multiple**.

Example 6

Find a common multiple of 8 and 12.

Multiples of 8:   8, 16, 24, 32, …
Multiples of 12: 12, 24, …

24 is a common multiple of 8 and 12.

List the multiples of each number.
The common multiples are the numbers that are in both lists.

Least common multiple

The **least common multiple** of two or more numbers is the *smallest* whole number that is a multiple of all of the numbers.

You can find the LCM by writing down multiples of each number.

'Least common multiple' is written as LCM.

Example 7

Find the LCM of 6, 8 and 10.

Multiples of 10: 10, 20, 30, …
Multiples of 6:   30, 60, 90, 120, 180, 210, 240, … ●————
Multiples of 8:   40, 80, 120, … ●

> Multiples of ten end in zero, so you only need list multiples of 6 and multiples of 8 that end in zero.

The LCM of 6, 8 and 10 is 120.

> 120 is the lowest number that all three numbers divide into exactly.

Exam practice 3E

1 Find a number that is a multiple of
   a 6 and 9   b 4 and 7   c 8 and 16   d 8 and 12.

2 Find a number that is a multiple of
   a 3, 4 and 5   b 4, 5 and 6. ●

> List the multiples of 3, 4 and 5 until you find a number that is in all three lists.

A01 3 a Kelly said that 140 is a multiple of both 5 and 7.
            Is Kelly correct? Give a reason for your answer.
         b Alistair said that 126 is a multiple of 7 and 9.
            Is Alistair correct? Give a reason for your answer.

> 140 is a multiple of 5 because 5 divides into 140 exactly. 140 is a multiple of 7 if 7 divides into 140 exactly.

4 a Work out the smallest number bigger than 54 that is a multiple of 6 and 9.
   b Calculate the largest number less than 100 that is a multiple of 8 and 5.

5 Find the least common multiple of
   a 2 and 3   b 6 and 15   c 3 and 4
   d 3 and 6   e 10 and 15   f 4 and 18
   g 18 and 24   h 12 and 15   i 15 and 18.

6 Find the least common multiple of
   a 9, 12 and 18   b 12, 16 and 24.

A01 7 Hetty said that the smallest number that 24 and 36 would divide into exactly is 144.
         Is Hetty correct? Give a reason for your answer.

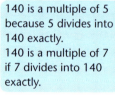

> The length must be a multiple of 3 to give an exact number of pieces that are 3 m long. The length must also be a multiple of 4 to give an exact number of pieces 4 m long. So you need to find the LCM of 3 and 4.

8 Find the smallest length of tape that can be cut into an exact number of pieces that are either 3 m long or 4 m long. ●

9 What is the smallest sum of money that can be made up of an exact number of
   a £20 notes or £50 notes   b £2 coins or £5 notes?

10 Find the shortest length of wire that can be cut into equal
     lengths that are either 4 cm or 5 cm or 10 cm long.

11 Two model trains travel round a track.
     One train takes 12 seconds to go round.
     The other train takes 15 seconds to go round.
     They start side by side.
     How many seconds will it be before they are side by side again?

12 Rectangular tiles measure 15 cm by 9 cm.
     What is the length of the side of the smallest square area that
     can be exactly covered with whole tiles?

13 Look at this list of numbers.
       3, 4, 7, 8, 12, 17, 21, 36, 40
     Write down
     a   a prime number larger than 10
     b   a factor of 14
     c   a square number
     d   a common multiple of 4 and 8
     e   a common factor of 12 and 21.

3.6 Index notation

Index notation can be used to simplify expressions such as
$2 \times 2 \times 2 \times 2$.

This number is called
the **index** or **power**.

So $2 \times 2 \times 2 \times 2 = 2^4$

This number is called
the **base**.

You say '2 to the power 4' or '2 to the 4'.

Example 8

Write $7 \times 7 \times 7 \times 7 \times 7$ using index notation.

$7 \times 7 \times 7 \times 7 \times 7 = 7^5$

There are five lots
of seven multiplied
together.

Example 9

Simplify $3 \times 5 \times 3 \times 5 \times 3$.

$3 \times 5 \times 3 \times 5 \times 3 = 3 \times 3 \times 3 \times 5 \times 5$

You can do the multiplication in any
order. Rearrange the calculations so that
the threes are together and the fives are
together.

$= 3^3 \times 5^2$

You cannot simplify $3^3 \times 5^2$ any further.

Example 10

Find the value of 7^3.

> 7^3 means $7 \times 7 \times 7$.

$7^3 = 7 \times 7 \times 7$

$\quad = 49 \times 7$

$\quad = 343$

When the power is 2 or 3, special names are used;
$\quad$ 5^2 is called 'five squared'
and $\quad$ 5^3 is called 'five cubed'.
You need to know the squares of the numbers 2 to 15.

Exam practice 3F

1 Write each of the following using index notation.
  a $2 \times 2 \times 2$ b $3 \times 3 \times 3 \times 3$
  c $5 \times 5 \times 5 \times 5$ d $7 \times 7 \times 7 \times 7 \times 7 \times 7$

2 Paula said '$3 \times 3 \times 3 \times 3 \times 3 \times 3$ is 3^5.'
  Is Paula correct? Give a reason for your answer.

3 Find the value of
  a 2^5 b 5^2 c 3^2 d 3^4.

> For part **a**, count the number of twos. Make sure you have 5 of them multiplied together.

4 a Which is the larger: 2^3 or 3^2?
  b Which is the smaller: 2^5 or 5^2?

A01

5 a Zoe said the value of 3^3 was 9.
      Is Zoe correct? Give a reason for your answer.
  b Bill said the value of 2^4 was 16.
      Is Bill correct? Give a reason for your answer.

> Start your list like this:
> $1^2 = 1$
> $2^2 = 4$
> $3^2 = 9$

6 Write down the first fifteen square numbers.

7 Write down the value of
  a 5 cubed   b 6 cubed   c 4 cubed   d 8 cubed.

> 5 cubed $= 5^3$
> $\quad = 5 \times 5 \times 5$

8 Simplify:
  a $2 \times 2 \times 7 \times 7$ b $3 \times 3 \times 3 \times 2 \times 2$
  c $2 \times 3 \times 3 \times 5 \times 2 \times 5$ d $7 \times 7 \times 7 \times 3 \times 5 \times 7 \times 3$
  e $13 \times 5 \times 13 \times 5 \times 13$ f $3 \times 5 \times 5 \times 3 \times 7 \times 3 \times 7$

9 Find the value of
  a $2^2 \times 3^3$ b $3^2 \times 5^2$
  c $2^2 \times 3^2 \times 5$ d $2 \times 3^2 \times 7$.

A01

10 Di wrote '$2^3 \times 5^2 = 10^5$'
  Di is wrong. Explain the mistake that Di made.

11 Find the value of   **a** $(2^3)^2$ **b** $(3^2)^2$.

12 Anne wrote $(3^2)^3 = 6^3$.
This is wrong. Explain the mistake that Anne made.

13 Write these numbers in full.
   a 10^2 **b** 10^3 **c** 10^5 **d** 10^6

14 Write these numbers as powers of 10.
   a 100   **b** 10   **c** 1 000 000   **d** 100 000 000

15 Write these numbers as powers of 2.
   a 2   **b** 16   **c** 32

16 Mushad said 'If you square a prime number, the answer is always an odd number.'
Give an example to show that Mushad is wrong.

17 Jon said 'When a and b are prime numbers, $a^3 + b^3$ is always an even number.'
Give an example to show that Jon is wrong.

18 Write True or False for each statement.
Give reasons your answers.
   a   The square of an even number is always even.
   b   The cube of an even number is always odd.
   c   The square of an odd number is always even.

> Remember that you work out the inside of a bracket first.

3.7 Square and cube roots

When a number is squared, the starting number is called a **square root** of the result.

Any positive number has two square roots, one positive and one negative.

> $4 = 2 \times 2$, so 2 is a square root of 4.
> $(-2) \times (-2)$ is also equal to 4, so -2 is also a square root of 4.

A negative number cannot have a square root because all square numbers are positive.

> You get a negative number only by multiplying a negative number and a positive number.

The symbol $\sqrt{}$ is used for 'square root',

$\sqrt{64}$ means the positive square root of 64, so $\sqrt{64} = 8$.

$-\sqrt{64}$ means the negative square root of 64, so $-\sqrt{64} = -8$.

A whole number whose square root is also a whole number is called a square number. It is also called a perfect square.

You should know the squares of the numbers from 2 to 15.

You can then write down their square roots.

Example 11

Find $\sqrt{196}$.

$\sqrt{196} = 14$

$14^2 = 196$ so $\sqrt{196} = 14.$

When a number is cubed, the starting number is called a **cube root** of the result.
4 is the cube root of 64 because $4^3 = 64$.

$8 = 2 \times 2 \times 2$ so 2 is the cube root of 8.

The symbol $\sqrt[3]{}$ is used for 'cube root'.

$\sqrt[3]{8}$ means the cube root of 8, so $\sqrt[3]{8} = 2$.

Each number has only one cube root.
If the number is positive the cube root is positive. $\sqrt[3]{27} = 3$

If the number is negative the cube root is negative. $\sqrt[3]{-8} = -2$

Not every whole number has an exact cube root.

You should know the cubes of 2, 3, 4, 5 and 10.

Example 12

Find $\sqrt[3]{125}$.

$\sqrt[3]{125} = 5$

$5^3 = 125$ so $\sqrt[3]{125} = 5.$

Exam practice 3G

1   Write down the positive square root of
   a   16   b   25   c   100
   d   81   e   169   f   144
   g   36   h   225

2   Dave said that one of the square roots of 256 is 17.
   Is he right? Explain your answer.

3   Find the square roots of 400.

You can write 400 as 4×100, so $\sqrt{400} = \sqrt{4} \times \sqrt{100}.$

4   Write down the smallest number that you need to multiply 8 by to give a perfect square.

5   Express 900 as the product of two square numbers.
   Use your answer to write down the value of $\sqrt{900}$.

Remember, 'product' means multiply.

6   Express 484 as a product of factors, each of which is a perfect square.
   Use your answer to write down the value of $\sqrt{484}$.

Remember that a perfect square is the same as a square number.

7 Work out:
   a $\sqrt{324}$ b $\sqrt{3600}$ c $\sqrt{625}$

8 Write down the cube root of   a 125   b -8 c 1000   d -27.

9 Express 8000 as a product of two numbers that have integer cube roots.
   Use your answer to find $\sqrt[3]{8000}$.

10 Express 216 as a product of two factors, each of which is the cube of a number you know.
    Use your answer to write down $\sqrt[3]{216}$.

11 Find the value of
    a $3^2 - \sqrt{4}$ b $\sqrt{49} + 5^3$ c $6^2 - \sqrt{225}$ d $\sqrt{64} - \sqrt[3]{64}$.

3.8 Multiplying and dividing numbers written in index notation

You can multiply together powers of the *same* number by adding the powers.

Example 13

Simplify $2^3 \times 2^2$.

$2^3 \times 2^2 = 2^{3+2} = 2^5$

> 2^3 and 2^2 are both powers of 2 so add the powers.

You can divide different powers of the *same* number by subtracting the powers.

Example 14

Simplify a $2^5 \div 2^2$ b $5^2 \div 5^2$.

a $2^5 \div 2^2 = 2^{5-2} = 2^3$
b $5^5 \div 5^2 = 5^{2-2} = 5^0 = 1$

> You are dividing so subtract the powers.

> $5^0 = 1$ because
> $5^2 \div 5^2 = 25 \div 25 = 1$.

Any number to the power of 0 is 1.

You cannot use the rules when the base numbers are different.
This means that you can not simplify $5^3 \times 3^2$ or $5^3 \div 3^2$.

Exam practice 3H

1 Write as a single expression in index form.
   a $3^5 \times 3^2$ b $7^5 \times 7^3$ c $9^2 \times 9^8$
   d $2^2 \times 2^7$ e $4^7 \times 4^9$ f $5^4 \times 5^4$

2 Graham said that $10^4 \times 10^3 = 10^{12}$.
   a Explain why Graham is wrong.
   b Write $10^4 \times 10^3$ as a single power of 10.

3 Simplify:

  a $4^5 \div 4^2$ b $2^7 \div 2^5$ c $5^6 \div 5^5$

  d $3^7 \div 3^4$ e $2^8 \div 2^6$ f $5^9 \div 5^2$

4 Simplify:

  a $12^4 \times 12^5$ b $7^{10} \div 7^6$ c $3^7 \div 3^2$

  d $4^3 \times 4$ e $5^5 \div 5^4$ f $3^4 \div 3$

> You can write 4 as 4^1 and 3 as 3^1.

5 Write $(2^3)^2$ as a single power of 2.

> $(2^3)^2 = 2^3 \times 2^3$

6 Write $(3^3)^4$ as a single power of 3.

7 Viv wrote that $(5^3)^4 = 5^7$.

  Is Viv correct? Explain your answer.

8 Find the value of

  a $3^3 \div 3^3$ b $4^3 \div 4^2$ c $7^3 \times 7^2 \div 7^5$ d $29^5 \div 29^5$.

3.9 Prime factors

When a factor of a number is a prime, it is called a **prime factor**.

Every number can be written as a **product** of prime factors.

Example 15

Express 48 as a product of its prime factors.

$48 = 4 \times 12$
$ = 2 \times 2 \times 3 \times 4$
$ = 2 \times 2 \times 3 \times 2 \times 2$
$ = 2 \times 2 \times 2 \times 2 \times 3 = 2^4 \times 3$

> Simplify your answer.

> Start by writing 48 as the product of any two factors: $48 = 4 \times 12$.
> Each factor that is not prime can be written as the product of two factors. Repeat this until all the factors are prime numbers.

A more organised approach helps for larger numbers.

Example 16

Express 2100 as a product of its prime factors.

$2100 \div \mathbf{2} = 1050$
$1050 \div \mathbf{2} = 525$
$525 \div \mathbf{3} = 175$
$175 \div \mathbf{5} = 35$
$35 \div \mathbf{5} = 7$
$7 \div \mathbf{7} = 1$

> Start by dividing by 2 as many times as possible.
> Then divide by 3 as many times as possible. Then divide by each prime number in turn until you are left with 1.

So $2100 = 2 \times 2 \times 3 \times 5 \times 5 \times 7$

$ = 2^2 \times 3 \times 5^2 \times 7$

You can use prime factors to find the HCF and the LCM of two or more numbers.

Example 17

Find the HCF and the LCM of 48, 52 and 104.

$48 = 2 \times 2 \times 2 \times 2 \times 3$

$52 = 2 \times 2 \times 13$

$104 = 2 \times 2 \times 2 \times 13$

The HCF is $2 \times 2 = 4$.

The LCM is $2 \times 2 \times 2 \times 2 \times 3 \times 13 = 624$.

> The HCF is the product of the prime factors that are common to 48, 52 and 104.

> To find the LCM, start with the prime factors of the smaller number then put in those factors of the larger number that are not already included.

Exam practice 3I

1 Write each number as a product of prime numbers.

a 36	b 78	c 525
d 264	e 100	f 315
g 240	h 300	i 154

2 Write each number as a product of its prime factors using index notation.

a 121	b 63	c 112
d 111	e 44	f 441

3 Write each number as a power of a prime number.

a 4	b 8	c 49
d 32	e 9	f 64

4 a Write 24 as a product of its prime factors.
   b Write 18 as a product of its prime factors.
   c Find the highest common factor of 24 and 18.
   d Find the least common multiple of 24 and 18.

5 a Find the highest common factor of 36, 78 and 108.
   b Find the least common multiple of 36, 78 and 108.

> Use your answers to question **1**, parts **a** and **b**.

Mini coursework task

These numbers are called triangular numbers: 1 3 6 10 15 21 28. They form a pattern.

Add 2 to the first number to get the second number,

add 3 to the second number to get the third number,

add 4 to the third number to get the fourth number, and so on.

- Write down the next two triangular numbers after 28.
- What do you need to add to the 20th number to get the 21st number?
- What other patterns can you find in the triangular numbers?
- Why do you think they are called triangular numbers?

Summary of key points

- 1 is not a prime number. 2 is the only even prime number.
- A factor of a number divides into that number exactly,
  e.g. 7 is a factor of 28 but 3 is not.
- A multiple of a number can be divided by that number exactly,
  e.g. 49 and 63 are both multiples of 7.
- The square of a number is found by multiplying the number by itself
  e.g. the square of 4 is $4 \times 4 = 16$.
- The cube of a number is found by multiplying a number by itself and then multiplying the result by itself e.g. the cube of 2 is $2 \times 2 \times 2 = 8$.
- A positive number has two square roots, one positive and one negative, e.g. the square roots of 25 are 5 and -5.
- 3^3 is a shorthand way of writing $3 \times 3 \times 3$.
- You can multiply together powers of the *same* number by adding the powers.
- You can divide powers of the *same* number by subtracting the powers.
- (any number)^0 = 1.

Most students who get GRADE E or above can:
- use index notation and find factors and multiples.

Most students who get GRADE C can also:
- multiply and divide numbers in index notation
- find the HCF and LCM of two or more numbers.

Glossary

Base	in index notation the number which is raised to a power, e.g. in the expression 3^6, the base is 3
Common factor	a number that will divide exactly into two or more numbers
Common multiple	a number which two or more numbers will divide into exactly
Counting numbers	the whole numbers 1, 2, 3, 4, ...
Cube root	the number which when cubed gives the original number, e.g. the cube root of 27 is 3
Even number	a number that divides exactly by 2
Factor	a number that will divide exactly into a given number
Highest common factor	the largest number that is a common factor of two or more numbers
Index (plural indices)	another word for power and it tells you how many of the base number to multiply together
Integer	a positive or negative whole number
Least common multiple	the smallest number that is a common multiple of two or more numbers
Multiple	a number that the given number divides into exactly
Natural numbers	another name for the counting numbers
Odd number	a number that does not divide exactly by 2
Power	tells you how many of the base number to multiply together
Product	the result of multiplying numbers together
Prime factor	a factor of a number that is also a prime number
Prime number	a number that has only 2 factors, itself and 1
Square numbers	a number that has a whole number as its square root
Square root	a number which when multiplied by itself gives the original number

4 Fractions

4.1 Fractions

A **fraction** is used to represent part of a quantity.

The top number is called the **numerator**. It tells you the number of parts.

The fraction 'three-quarters' is written as $\frac{3}{4}$.

The bottom number is called the **denominator**. It tells you how many equal-sized parts the whole has been divided into.

Did you know
that showing a fraction by writing one number above the other, like $\frac{3}{4}$, was probably introduced by a Hindu mathematician name Brahmagupta?

Example 1

Shade $\frac{3}{5}$ of this shape:

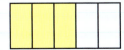

The bar is divided into 5 equal sized parts so shade 3 of them.

Exam practice 4A

1 Write down the numerator of each fraction.

 a $\frac{1}{4}$ b $\frac{3}{5}$ c $\frac{2}{7}$ d $\frac{5}{12}$

2 Write down the denominator of each fraction.

 a $\frac{2}{5}$ b $\frac{1}{4}$ c $\frac{5}{6}$ d $\frac{11}{12}$

3 Write down

 a the numerator of $\frac{7}{12}$ b the denominator of $\frac{7}{12}$

 c the numerator of $\frac{7}{8}$ d the denominator of $\frac{7}{8}$

4 Write each fraction in figures.

 a one-third   b four-fifths   c five-eighths

 d three-quarters   e seven-twelfths   f seven-tenths

5 Write down the fraction with

 a numerator 3 and denominator 7

 b denominator 11 and numerator 8

 c numerator 9 and denominator 12

 d denominator 20 and numerator 5.

6 Write down the fraction of each shape that is shaded.

 a   b   c   d

  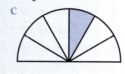

 e   f   g   h

 7 Copy each shape and shade the fraction asked for.

 a $\frac{5}{6}$ b $\frac{1}{4}$

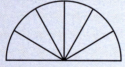 

 c $\frac{5}{8}$ d $\frac{7}{12}$

 e $\frac{3}{8}$

8 Jenny says that $\frac{1}{3}$ of this triangle is shaded. Explain why she is wrong.

4.2 Equivalent fractions

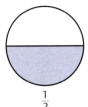 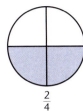 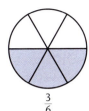 

$\frac{1}{2}$ $\frac{2}{4}$ $\frac{3}{6}$ $\frac{4}{8}$

From the diagrams you can see that $\frac{1}{2}, \frac{2}{4}, \frac{3}{6}$ and $\frac{4}{8}$ of the circle are all the same size.

The fractions $\frac{1}{2}, \frac{2}{4}, \frac{3}{6}$ and $\frac{4}{8}$ are called **equivalent fractions**.

You get an equivalent fraction when you multiply or divide the top and the bottom of a fraction by the same number.

Example 2

Write $\frac{1}{5}$ as an equivalent fraction with 15 as the denominator.

$$\frac{1}{5} = \frac{3}{15}$$

> You need to write $\frac{1}{5}$ as $\frac{\Box}{15}$
> $5 \times 3 = 15$, so multiply top and bottom by 3.

When the numerator and denominator are divided by the same number, you get an equivalent fraction with a smaller numerator and denominator.
This is called **simplifying** or **cancelling**.

When the fraction has been simplified to give the smallest possible numerator and denominator, it is in its **lowest possible terms**.

Example 3

Simplify $\frac{15}{75}$.

$$\frac{15}{75} = \frac{15}{75} = \frac{1}{5}$$

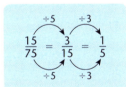

$$\frac{15}{75} = \frac{3}{15} = \frac{1}{5}$$

Exam practice 4B

1 Write each fraction as a number of eighths.

   a $\frac{1}{2}$ b $\frac{1}{4}$ c $\frac{3}{4}$

> To write $\frac{1}{2}$ as a number of eighths, you need to multiply top and bottom by the number that changes 2 to 8.

2 Write each fraction as a number of twelfths.

   a $\frac{1}{2}$ b $\frac{2}{3}$ c $\frac{3}{4}$ d $\frac{1}{6}$

3 Find equivalent fractions by filling in the missing numbers.

   a $\frac{2}{3} = \frac{4}{\Box}$ b $\frac{3}{4} = \frac{\Box}{12}$ c $\frac{1}{2} = \frac{5}{\Box}$

> You multiply the numerator of $\frac{2}{3}$ by 2 to get 4, so you must also multiply the denominator of $\frac{2}{3}$ by 2 to find the missing number.

4 Write each fraction as an equivalent fraction with a denominator of 15.

   a $\frac{2}{5}$ b $\frac{1}{3}$ c $\frac{2}{3}$

5 Write each fraction as an equivalent fraction with a denominator of 24.

   a $\frac{1}{3}$ b $\frac{1}{2}$ c $\frac{1}{4}$ d $\frac{3}{4}$ e $\frac{2}{3}$ f $\frac{1}{8}$

6 Write each fraction as an equivalent fraction with a denominator of 32.

   a $\frac{3}{8}$ b $\frac{1}{16}$ c $\frac{3}{4}$

7 Write each fraction as an equivalent fraction with a denominator of 45.

   a $\frac{2}{15}$ b $\frac{4}{9}$ c $\frac{2}{3}$

8 Complete these equivalent fractions.

   a $\frac{2}{5} = \frac{\square}{10} = \frac{6}{\square} = \frac{\square}{25}$ b $\frac{3}{4} = \frac{\square}{8} = \frac{9}{\square} = \frac{\square}{20}$ c $\frac{8}{20} = \frac{\square}{5} = \frac{4}{\square} = \frac{12}{\square}$

9 Give each fraction in its simplest form.

   a $\frac{3}{12}$ b $\frac{8}{16}$ c $\frac{4}{8}$ d $\frac{9}{12}$ e $\frac{15}{20}$ f $\frac{10}{15}$

   g $\frac{8}{12}$ h $\frac{6}{8}$ i $\frac{12}{20}$ j $\frac{75}{100}$ k $\frac{80}{100}$ l $\frac{24}{30}$

   m $\frac{9}{36}$ n $\frac{12}{18}$ p $\frac{9}{27}$ q $\frac{16}{32}$ r $\frac{40}{80}$ s $\frac{24}{32}$

> This means cancel the fraction as far as you can.

10 Write down the fraction that is shaded. Simplify your answers if possible.

   a   b   c

11 Which two of these rectangles have the same fraction shaded?

   A   B   C

4.3 Comparing fractions

To compare fractions with different denominators change them into equivalent fractions with the same denominator.

Example 4

Which is the smaller $\frac{2}{9}$ or $\frac{1}{6}$?

$$\frac{2}{9} = \frac{4}{18}$$

and $\frac{1}{6} = \frac{3}{18}$

so $\frac{1}{6}$ is smaller.

> Write both fractions as equivalent fractions with the same denominator.
>
> $3 < 4$ so $\frac{3}{18} < \frac{4}{18}$

Exam practice 4C

1 Write $\frac{5}{7}$ and $\frac{2}{3}$ as equivalent fractions with a denominator of 21.
   Which fraction is the larger ?

2 Find equivalent fractions for $\frac{2}{3}$ and $\frac{4}{5}$ that have the same denominator.
   Which fraction is the smaller?

3 a Which is the larger: $\frac{2}{9}$ of a cake or $\frac{1}{7}$ of it?
   b Which is the smaller: $\frac{2}{7}$ of a box of sweets or $\frac{3}{8}$ of it?
   c Which is the larger: $\frac{4}{5}$ of a loaf of bread or $\frac{6}{7}$ of it?

> You need to change $\frac{2}{9}$ and $\frac{1}{7}$ into equivalent fractions with the same denominator.

A01 4 Paul said that $\frac{5}{8}$ was smaller than $\frac{3}{4}$.
   Is Paul correct? Give a reason for your answer.

4.4 Mixed numbers and improper fractions

Fractions that are less than a whole unit are called **proper fractions**.

In the diagram there are one and a half circles or three half-circles.

One and a half is written as $1\frac{1}{2}$ and is called a **mixed number**.

Three halves is written as $\frac{3}{2}$ and is called an **improper fraction**.

To change a mixed number to an improper fraction, multiply the whole number by the denominator and add the result to the numerator.

> A mixed number contains a whole number and a fraction.

> In an improper fraction the numerator is greater than the denominator.

Example 5

Write $2\frac{1}{3}$ as an improper fraction.

$$2\frac{1}{3} = \frac{6+1}{3} = \frac{7}{3}$$

> $2 = \frac{6}{3}$

To change an improper fraction to a mixed number, divide the numerator by the denominator to give the number of units; the remainder is the number of fractional parts.

Example 6

Write $\frac{13}{6}$ as a mixed number.

$$\frac{13}{6} = 2 + \frac{1}{6} = 2\frac{1}{6}$$

$13 \div 6 = 2$ remainder 1.

Exam practice 4D

1 Write each improper fraction as a mixed number.

   a $\frac{3}{2}$ **b** $\frac{4}{3}$ **c** $\frac{5}{3}$ **d** $\frac{7}{4}$ **e** $\frac{12}{5}$ **f** $\frac{9}{4}$

   g $\frac{10}{3}$ **h** $\frac{11}{3}$ **i** $\frac{15}{4}$ **j** $\frac{11}{4}$ **k** $\frac{7}{3}$ **l** $\frac{11}{5}$

Remember that $3 \div 2$ and $\frac{3}{2}$ mean the same thing.

2 Write each mixed number as an improper fraction.

   a $1\frac{1}{4}$ **b** $2\frac{1}{3}$ **c** $3\frac{1}{4}$ **d** $2\frac{2}{5}$

   e $2\frac{3}{4}$ **f** $1\frac{2}{5}$ **g** $3\frac{3}{4}$ **h** $4\frac{1}{4}$

3 Work out each division. Give your answers as mixed numbers.

   a $5 \div 4$ **b** $10 \div 3$ **c** $8 \div 3$ **d** $12 \div 5$

   e $16 \div 5$ **f** $18 \div 7$ **g** $12 \div 7$ **h** $9 \div 5$

   i $10 \div 7$ **j** $11 \div 9$ **k** $12 \div 11$ **l** $11 \div 7$

4.5 Finding fractions

To find one quantity as a fraction of another,
- write both quantities in the same unit
- write the first quantity as the numerator and the second quantity as the denominator
- simplify the fraction.

You cannot find 15 minutes as a fraction of £1 because they are not the same kind of quantity and cannot be given in the same units.

Example 7

Find 2 eggs as a fraction of 6 eggs.

$$\frac{2}{6} = \frac{1}{3}$$

Example 8

Find 3 days as a fraction of 1 week.

1 week = 7 days

Write both quantities in the same unit.

So 1 day is $\frac{1}{7}$ of a week and 3 days is $\frac{3}{7}$ of a week.

Example 9

Find 15 minutes as a fraction of $1\frac{1}{2}$ hours.

$1\frac{1}{2}$ hours = 90 minutes,

so 15 minutes is $\frac{15}{90}$ of $1\frac{1}{2}$ hours.

$\frac{15}{90} = \frac{1}{6}$, so 15 minutes is $\frac{1}{6}$ of $1\frac{1}{2}$ hours.

(÷15 ... ÷15)

Exam practice 4E

1   Len took 4 chocolates from a box of 12 chocolates.
    What fraction was this?

2   Sally was driven 180 miles. She slept for 120 miles.
    What fraction of the journey was she asleep?

3   9000 people went to a concert. 4500 of them were under
    18 years old.
    What fraction was this?

4   There are 36 cars in the school car park. 8 of them are red.
    What fraction of the cars in the school car park are
    a red   b not red?

5   Find a   36 cm as a fraction of 144 cm
          b   60 cm as a fraction of 3 metres
          c   50p as a fraction of £5.

6   Each week, Jon saved £1.50 of his £4 pocket money.
    What fraction of his pocket money did Jon save?

7   What fraction of one hour is
    a   1 minute   b 10 minutes   c 45 minutes   d 36 minutes?

8   Clare has a 25 hectare field. She plants 15 hectares with wheat.
    What fraction of the field is planted with wheat?

9   When she went on holiday, Carol's luggage weighed 15 kg.
    When she got home her luggage weighed 18 kg.
    What fraction of this weight did her luggage originally weigh?

10 This pumpkin weighed 6 kg before it was carved.
    It now weighs 2 kg.
    What fraction of its weight was removed?

11 Simon gets £4.80 a week pocket money. He spends £2.70.
    What fraction of his pocket money is left?

12 In a quiz Pat scored 20 out of 80 on Section A and 40 out of 60
    on Section B.
    What fraction of the total score did Pat get?

Give your answers as fractions in their simplest form.

1 metre = 100 cm.

Change £5 to pence.

4.6 Addition and subtraction

Fractions with the same denominator can be added or subtracted by adding or subtracting their numerators.

Example 10

Work out **a** $\frac{2}{5} + \frac{1}{5}$ **b** $\frac{5}{8} - \frac{3}{8}$.

a $\frac{2}{5} + \frac{1}{5} = \frac{2+1}{5} = \frac{3}{5}$

b $\frac{5}{8} - \frac{3}{8} = \frac{5-3}{8} = \frac{2}{8} = \frac{1}{4}$

$\frac{2}{8}$ can be simplified by dividing top and bottom by 2.

You can add or subtract fractions with different denominators by writing them as equivalent fractions with the same denominator.

Example 11

Find $\frac{3}{4} - \frac{2}{3}$.

$\frac{3}{4} - \frac{2}{3} = \frac{9}{12} - \frac{8}{12}$

$= \frac{1}{12}$

Write both fractions with the same denominator.
Choose the least common multiple of the two denominators.

Exam practice 4F

1 Work out:
   a $\frac{1}{4} + \frac{1}{4}$ **b** $\frac{5}{8} + \frac{1}{8}$ **c** $\frac{7}{8} - \frac{3}{8}$
   d $\frac{7}{10} - \frac{3}{10}$ **e** $\frac{7}{12} + \frac{5}{12}$ **f** $\frac{7}{12} - \frac{5}{12}$

2 Find $\frac{1}{2} + \frac{1}{4}$ as a fraction in its lowest terms.

3 Work out: **a** $1 - \frac{3}{10}$ **b** $1 - \frac{5}{8}$ **c** $1 - \frac{4}{5}$

4 Find:
   a $\frac{1}{4} + \frac{3}{8}$ **b** $\frac{5}{12} + \frac{1}{2}$ **c** $\frac{5}{8} + \frac{1}{4}$ **d** $\frac{7}{12} + \frac{1}{6}$
   e $\frac{1}{2} - \frac{1}{4}$ **f** $\frac{1}{2} - \frac{3}{8}$ **g** $\frac{7}{8} - \frac{3}{4}$ **h** $\frac{3}{4} - \frac{3}{8}$

5 Work out:
   a $\frac{2}{5} + \frac{1}{4}$ **b** $\frac{1}{4} + \frac{7}{10}$ **c** $\frac{7}{10} - \frac{1}{2}$ **d** $\frac{2}{3} - \frac{1}{2}$ **e** $\frac{1}{3} + \frac{1}{5}$

6 Find $\frac{3}{4} - \frac{5}{8} + \frac{1}{2}$ as a fraction in its simplest form.

Give your answers as fractions in their lowest terms.

Both denominators will divide into 4 so change $\frac{1}{2}$ into a fraction with 4 as the denominator.

$1 = \frac{10}{10}$

Find the least common multiple of the denominators. Change the fractions to equivalent fractions with this number as the denominator.

4.7 Adding and subtracting with mixed numbers

You can add and subtract mixed numbers by changing them into improper fractions.

Example 12

Find   **a** $2\frac{1}{2} + 1\frac{3}{5}$ **b** $2\frac{1}{2} - 1\frac{3}{5}$.

a $2\frac{1}{2} + 1\frac{3}{5} = \frac{5}{2} + \frac{8}{5}$ ● ——— Change each mixed number to an improper fraction.

$= \frac{25}{10} + \frac{16}{10}$

$= \frac{41}{10} = 4\frac{1}{10}$ ● ——— Write your answer as a mixed number.

b $2\frac{1}{2} - 1\frac{3}{5} = \frac{5}{2} - \frac{8}{5}$

$= \frac{25}{10} - \frac{16}{10}$

$= \frac{9}{10}$

Exam practice 4G

1 Find:

   a $2\frac{1}{4} + 1\frac{1}{2}$ b $3\frac{3}{4} + 1\frac{1}{4}$ c $4\frac{3}{8} + 2\frac{1}{8}$

   d $2\frac{3}{4} - 1\frac{1}{2}$ e $4\frac{5}{8} - 2\frac{1}{4}$ f $3\frac{7}{8} - 2\frac{1}{2}$

> Write your answers as mixed numbers or fractions in their lowest terms.

2 Work out:

   a $2\frac{1}{4} + \frac{4}{5}$ b $3\frac{1}{4} + 1\frac{1}{5}$ c $2\frac{1}{2} + 1\frac{3}{4}$

   d $1\frac{1}{3} - \frac{3}{4}$ e $3\frac{1}{3} - 1\frac{1}{2}$ f $2\frac{1}{4} - 1\frac{2}{3}$

3 Stan has $5\frac{1}{4}$ metres of rope. Dilly buys $2\frac{3}{4}$ metres of it. How much does Stan have left?

4 Tim has two lengths of skirting board.
   One is $3\frac{1}{4}$ metres long and the other is $4\frac{7}{8}$ metres long.
   a How much skirting board does he have in total?
   b He needs $7\frac{5}{8}$ metres. How much will he have left over?

5 The ceiling of a room is $8\frac{1}{2}$ feet high.
   A picture, which is $2\frac{1}{4}$ feet high, hangs on a wall.
   The bottom of the picture is $4\frac{1}{2}$ feet from the floor.
   Work out the distance from the top of the picture to the ceiling.

Mini coursework task

You cannot use improper fractions in this activity.

Look at the digits 2, 3, 4, 5, 6 and 7.

- Use two of these digits to make a fraction that is
  i as large as possible   **ii** as small as possible.
- Use two of the digits on the top and two on the bottom to make a fraction that is
  i as large as possible   **ii** as small as possible.
- Use two pairs of these digits to make two equivalent fractions.
- Make two fractions with a difference that is
  i as large as possible   **ii** as small as possible.

> You can use digits to make fractions such as $\frac{2}{3}$ or $\frac{3}{7}$.

> No digit can be used more than once in the same fraction, so $\frac{23}{27}$ is not allowed because 2 is used twice.

> Each digit can be used only once. The numerator and denominator must have the same number of digits.

4.8 Multiplication

To multiply fractions, multiply the numerators together and multiply the denominators together.

Example 13

Calculate   **a** $\frac{2}{7} \times \frac{3}{5}$ **b** $\frac{3}{4} \times \frac{16}{21}$.

a $\dfrac{2}{7} \times \dfrac{3}{5} = \dfrac{6}{35}$

b $\dfrac{3}{4} \times \dfrac{16}{21} = \dfrac{{}^{1}\cancel{3} \times \cancel{16}^{4}}{{}_{1}\cancel{4} \times \cancel{21}_{7}}$

$ = \dfrac{1 \times 4}{1 \times 7}$

$ = \dfrac{4}{7}$

> You can cancel common factors before you multiply the numerators and denominators.

Exam practice 4H

Write your answers for each calculation as fractions in their simplest form.

1 $\frac{2}{3} \times \frac{1}{7}$

2 $\frac{4}{5} \times \frac{3}{5}$

3 $\frac{5}{12} \times \frac{4}{7}$

4 $\frac{2}{5} \times \frac{3}{4}$

5 $\frac{16}{21} \times \frac{3}{8}$

6 $\frac{4}{9} \times \frac{33}{44}$

7 $\frac{4}{5} \times \frac{10}{16}$

8 $\frac{3}{7} \times \frac{28}{33}$

9 $\frac{11}{27} \times \frac{18}{22}$

10 $\frac{4}{21} \times \frac{7}{8}$

11 $\frac{3}{8} \times \frac{21}{30} \times \frac{5}{7}$

12 $\frac{8}{9} \times \frac{4}{5} \times \frac{15}{16}$

13 $\frac{3}{4} \times \frac{7}{12} \times \frac{18}{21}$

14 $\frac{5}{8} \times \frac{12}{25} \times \frac{2}{9}$

4.9 Multiplying whole numbers and mixed numbers

You can multiply mixed numbers by changing them to improper fractions.

> A whole number can be written as a fraction by writing it over 1, so $4 = \frac{4}{1}$.

Example 14

Calculate   **a** $3\frac{1}{4} \times \frac{3}{13}$ **b** $8\frac{3}{4} \times 2\frac{2}{7}$ **c** $3 \times 6\frac{1}{9}$.

a $3\frac{1}{4} \times \frac{3}{13} = \frac{\overset{1}{\cancel{13}} \times 3}{4 \times \cancel{13}_{1}}$

$\qquad = \frac{3}{4}$

> $3\frac{1}{4} = \frac{12 + 1}{4} = \frac{13}{4}$

b $8\frac{3}{4} \times 2\frac{2}{7} = \frac{\overset{5}{\cancel{35}} \times \overset{4}{\cancel{16}}}{\underset{1}{\cancel{4}} \times \cancel{7}_{1}}$

$\qquad = \frac{5 \times 4}{1 \times 1}$

$\qquad = \frac{20}{1}$

$\qquad = 20$

c $3 \times 6\frac{1}{9} = \frac{3}{1} \times \frac{55}{9} = \frac{\overset{1}{\cancel{3}} \times 55}{1 \times \cancel{9}_{3}}$

$\qquad = \frac{55}{3}$

$\qquad = 18\frac{1}{3}$

> Write 3 as $\frac{3}{1}$.

Exam practice 4I

> Write your answers as mixed numbers.

1 Calculate: **a** $2\frac{1}{2} \times \frac{2}{5}$ **b** $3\frac{3}{4} \times \frac{3}{10}$ **c** $5\frac{2}{3} \times \frac{30}{34}$ **d** $2\frac{1}{5} \times \frac{2}{22}$

2 Calculate: **a** $\frac{7}{12} \times 2\frac{2}{5}$ **b** $8\frac{1}{3} \times 3\frac{3}{5}$ **c** $5\frac{1}{2} \times \frac{9}{11}$ **d** $2\frac{2}{7} \times 8\frac{3}{4}$

3 Calculate: **a** $4 \times 3\frac{3}{8}$ **b** $3\frac{1}{8} \times 16$ **c** $2\frac{2}{7} \times 14$ **d** $3\frac{3}{5} \times 10$

4 A rectangular paving stone measures $\frac{3}{8}$ m by $\frac{5}{8}$ m. What is the area of the paving stone?

> Area of a rectangle is length × breadth.

4.10 Reciprocals

> $\frac{1}{4}$ is the reciprocal of 4 and 4 is the reciprocal of $\frac{1}{4}$ because $\frac{1}{4} \times 4 = \frac{1}{4} \times \frac{4}{1} = 1$.

If the **product** of two numbers is 1 then each number is called the **reciprocal** of the other.

The reciprocal of a fraction is found by turning the fraction upside down.
The reciprocal of a number is 1 divided by that number.
0 does not have a reciprocal because you cannot divide by 0.

> So $\frac{4}{3}$ is the reciprocal of $\frac{3}{4}$.

Example 15

Find the reciprocal of **a** 5   **b** $\frac{7}{9}$.

   a The reciprocal of 5 is $\frac{1}{5}$.

                                                      $1 \div 5 = \frac{1}{5}$

   b The reciprocal of $\frac{7}{9}$ is $\frac{9}{7}$ or $1\frac{2}{7}$.

Exam practice 4J

Write down the reciprocals of the following numbers.

1 6   **2** $\frac{1}{2}$ **3** $\frac{1}{8}$ **4** $\frac{5}{9}$ **5** $\frac{5}{8}$

6 20   **7** 100   **8** $\frac{12}{7}$ **9** $\frac{5}{11}$ **10** $\frac{20}{17}$

4.11 Division by a fraction

To divide by a fraction turn the fraction upside down and multiply by it.

> This is the same as multiplying by the reciprocal.

Example 16

Find the value of   **a** $24 \div \frac{3}{4}$ **b** $\frac{8}{75} : \frac{4}{15}$ **c** $5\frac{5}{8} \div 6\frac{1}{4}$ **d** $2\frac{1}{2} \div 5$.

  a $24 \div \frac{3}{4} = \frac{24}{1} \div \frac{3}{4} = \frac{\overset{8}{\cancel{24}} \times 4}{1 \times \cancel{3}_1}$

           $= 32$

> Write 24 as $\frac{24}{1}$. Turn $\frac{3}{4}$ upside down and multiply.

  b $\frac{8}{75} \div \frac{4}{15} = \frac{\overset{2}{\cancel{8}} \times \overset{1}{\cancel{15}}}{\underset{5}{\cancel{75}} \times \cancel{4}_1}$

           $= \frac{2}{5}$

> Turn $\frac{4}{15}$ upside down and then multiply.

  c $5\frac{5}{8} \div 6\frac{1}{4} = \frac{45}{8} \div \frac{25}{4}$

          $= \frac{\overset{9}{\cancel{45}} \times \overset{1}{\cancel{4}}}{\underset{2}{\cancel{8}} \times \cancel{25}_5}$

          $= \frac{9}{10}$

> Write the mixed numbers as improper fractions.

  d $2\frac{1}{2} \div 5 = \frac{5}{2} \div \frac{5}{1}$

         $= \frac{\overset{1}{\cancel{5}}}{2} \times \frac{1}{\cancel{5}_1} = \frac{1}{2}$

> Write $2\frac{1}{2}$ as an improper fraction and write 5 as $\frac{5}{1}$.

Exam practice 4K

1 **a** How many halves are there in 6?
    b How many times does $\frac{1}{7}$ go into 5?

2 Find: **a** $18 \div \frac{6}{11}$ **b** $15 \div \frac{5}{9}$ **c** $35 \div \frac{5}{7}$

         d $9 \div \frac{3}{13}$ **e** $\frac{3}{4} \div 30$ **f** $\frac{4}{5} \div 20$

3   Find:
   a $\frac{8}{21} \div \frac{2}{7}$
   b $\frac{35}{42} \div \frac{5}{6}$

   c $\frac{15}{22} \div \frac{5}{11}$
   d $\frac{3}{28} \div \frac{9}{14}$

4   Calculate:
   a $8\frac{3}{4} \div 12\frac{1}{2}$
   b $9\frac{3}{4} \div 1\frac{5}{8}$

   c $3\frac{3}{10} \div 8\frac{4}{5}$
   d $6\frac{3}{4} \div 7$

5   Divide:
   a $1\frac{11}{21}$ by $9\frac{1}{7}$
   b $10\frac{5}{6}$ by $6\frac{1}{2}$

6   a   How many pieces of string $\frac{5}{8}$ m long can be cut from a piece that is $12\frac{4}{5}$ m long?
   b   How much is left over?

7   James read 20 pages of a book in $\frac{1}{2}$ hour.
   a   What fraction of an hour did it take him to read one page?
   b   How many minutes did it take him to read one page?

4.12 Fractions of a quantity

To find a fraction of a quantity, you divide by the denominator, then multiply the answer by the numerator.

Example 17

Find $\frac{2}{3}$ of the number of tiles on this floor.

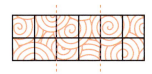

There are 12 tiles.

$\frac{1}{3}$ of 12 = 12 ÷ 3 = 4.

So $\frac{2}{3}$ of 12 = 4 × 2 = 8.

A third means one of 3 equal sized parts. Divide these twelve tiles into 3 equal parts. You want two of these parts so multiply by 2.

Example 18

Find $\frac{5}{8}$ of £48.

$\frac{1}{8}$ of £48 = £48 ÷ 8 = £6.

So $\frac{5}{8}$ of £48 = 5 × £6 = £30.

First find $\frac{1}{8}$, then multiply your answer by 5.

Exam practice 4L

1   a   Copy this shape and shade $\frac{5}{8}$.

Find the number of small triangles that are $\frac{1}{8}$ of the whole shape.

b Copy this shape and shade $\frac{3}{4}$.

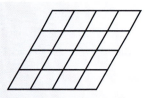

c Copy this shape and shade $\frac{1}{2}$.

2 Find: **a** $\frac{1}{4}$ of £24   **b** $\frac{2}{7}$ of 14 days

          c $\frac{1}{3}$ of 24 hours   **d** $\frac{2}{3}$ of 60 seconds.

3 Find: **a** $\frac{1}{5}$ of £35   **b** $\frac{3}{4}$ of 36 cm

          c $\frac{1}{9}$ of 27 m   **d** $\frac{5}{8}$ of 36 ft.

4 Find: **a** $\frac{2}{3}$ of 48 euros   **b** $\frac{3}{5}$ of a year of 365 days

          c $\frac{2}{5}$ of 85 kg   **d** $\frac{3}{4}$ of 96 pence.

> Remember to give the units.

5 Find: **a** $\frac{3}{8}$ of 88 miles   **b** $\frac{7}{16}$ of 48 litres

          c $\frac{4}{9}$ of 63 kilometres   **d** $\frac{7}{8}$ of 1 day.

> There are 24 hours in a day.

6 Find: **a** $\frac{3}{5}$ of 35 p   **b** $\frac{4}{7}$ of 49 cm

          c $\frac{4}{5}$ of 1 year of 365 days   **d** $\frac{7}{12}$ of 1 hour.

7 How many more squares must you shade so that $\frac{3}{4}$ of this shape is shaded?

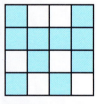

8 How many more squares must you shade so that $\frac{2}{3}$ of this rectangle is shaded?

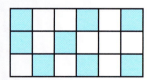

9 The full price of a train ticket is £64.
If you book a week before the journey, you get a $\frac{1}{4}$ off this price.
Work out how much you save if you book a week ahead.

10 Frank earns £327 a week.
He loses $\frac{1}{3}$ of this on deductions for tax, national insurance and pension contributions.
What sum of money is deducted from Frank's pay?

11 A cottage costs £450 to rent for a week in high season.
The rent in low season is $\frac{2}{3}$ of this.
Find the rent for a week in low season.

12 A holiday in Spain for an adult costs £650.
    The cost for a teenager is $\frac{9}{10}$ the cost for an adult.
    Work out the cost of the holiday for a teenager.

13 In a sale prices are reduced by $\frac{1}{4}$.
    The full price of a pair of jeans is £28.
    a Work out the reduction.
    b Find the sale price.

14 A garage owner takes $\frac{1}{10}$ off a bill if it is paid straight away.
    Steve's bill is £140.
    a Work out the reduction.
    b How much must Steve pay?

15 An empty van has a total weight of 2000 kg.
    When the van is fully loaded, the total weight is $3\frac{1}{2}$ times its empty
    weight.
    a Work out the total weight of the fully loaded van.
    b What is the weight of the load in the van?

16 The cost of a rail ticket is reduced by a third with a rail card.
    The full cost of a ticket is £96.
    Find the cost of this ticket with a rail card.

17 Sally had 60 metres of string.
    She used $\frac{3}{5}$ of it on Monday and $\frac{1}{4}$ of it on Tuesday.
    a What length of string did she use
       i on Monday   ii on Tuesday?
    b What length remained?

18 Claire did a 12 kilometre cross country trek.
    She ran $\frac{1}{4}$ of the way.
    a How far did she still have to go?
    She walked for $\frac{2}{3}$ of the remaining distance
    before she stopped for a rest.
    b How far did she walk before she rested?
    c How far did she still have to go?

19 In an election, Peter got $\frac{5}{12}$ of the votes and Sue got $\frac{2}{5}$ of the votes.
    60 people voted.
    a How many voted for   i Peter   ii Sue?
    b How many did not vote for Peter or for Sue?

20 Anne, Julian and Cheryl need £5000 to open a hairdressing
    salon.
    Anne contributes $\frac{1}{2}$, Julian $\frac{1}{5}$ and Cheryl the remainder.
    How much money does each person contribute?

This symmetrical mosaic
is in the Alhambra in
Spain. It was made about
700 years ago.

Summary of key points

- You can find equivalent fractions by multiplying the numerator and denominator by the same number.
- You can simplify fractions by dividing the numerator and denominator by the same number.
- You can add and subtract fractions by changing them to equivalent fractions with the same denominator.
- You can find one quantity as a fraction of another by first making sure that they are in the same units, then placing the first quantity over the second.
- $15 \div 4$ and $\frac{15}{4}$ mean the same thing.
- You can compare the size of fractions by changing them to equivalent fractions with the same denominator and then comparing their numerators.
- You can multiply two fractions together by multiplying their numerators together and multiplying their denominators together.
- Convert mixed numbers into improper fractions before carrying out calculations.
- To divide by a fraction turn it upside down and multiply by it.
- To find a fraction of a quantity divide by the denominator and multiply by the numerator.

Most students who get GRADE E or above can:
- find equivalent fractions
- add and subtract fractions
- multiply a fraction by a whole number
- find a fraction of a quantity.

Most students who get GRADE C can also:
- multiply and divide with fractions and mixed numbers.

Glossary

Cancel	find a simpler equivalent fraction by dividing the numerator and denominator by the same number
Denominator	the bottom number in a fraction
Equivalent fraction	an equal fraction with a different numerator and denominator
Fraction	part of a quantity
Improper fraction	a fraction where the numerator is larger than the denominator
Lowest possible terms	a fraction that has been simplified as far as possible
Mixed number	a number that contains a whole number part and a fraction part, e.g. $3\frac{1}{3}$
Numerator	the top number in a fraction
Product	the result when two numbers are multiplied together
Proper fraction	a fraction that is less than 1 whole unit, so the numerator is smaller than the denominator
Reciprocal	the number which when multiplied by the original number gives an answer of 1
Simplifying a fraction	dividing the numerator and denominator by the same number to get an equivalent fraction with a smaller numerator and denominator

5 Decimals

This chapter will show you:
- ✓ how to use place value diagrams
- ✓ how to write a decimal as a fraction
- ✓ how to add or subtract decimal numbers
- ✓ how to multiply or divide a decimal number by a whole number or another decimal number
- ✓ how to change a fraction to a decimal
- ✓ that some fractions form recurring decimals
- ✓ the meaning of standard form

Before you start you need to know:
- ✓ how to add and subtract whole numbers
- ✓ how to do short and long division
- ✓ the meaning of powers of ten

5.1 Decimal places

The position of a digit in a number is called its **place value**. It tells you the value of that digit.

Mixed numbers can be written as decimal numbers. They are written with a decimal point after the units.

1000s	100s	10s	units •	$\frac{1}{10}$ s	$\frac{1}{100}$ s
		5	0 •	7	1

This column represents hundredths. It is called the **second decimal place**.

The number is 50.71. In words you write fifty point seven one.

This is the decimal point.

This column represents tenths. It is called the **first decimal place**.

Example 1

Write nought point nought five in figures.

0.05

You can use a place value diagram to help. There are zero units and zero tenths.

units •	$\frac{1}{10}$ ths	$\frac{1}{100}$ ths	$\frac{1}{1000}$ ths
0 •	0	5	

Example 2

Write 2.503 in words.

Two point five nought three.

units •	$\frac{1}{10}$ ths	$\frac{1}{100}$ ths	$\frac{1}{1000}$ ths
2 •	5	0	3

Comparing decimals

You can compare the size of two or more decimals by looking at the figures in each place value.

Example 3

Which is larger, 4.67 or 4.632?

4.67 is larger than 4.632.

> Look first at the number of units - they are the same. Next look at the number of tenths - they are also the same. Finally look at the number of hundredths - 7 is larger than 3, so 4.67 is larger than 4.632. Just because 4.632 has more digits than 4.67 does not mean 4.632 is larger than 4.67.

Exam practice 5A

1 Write these decimals in figures.
   a four point three
   b eight point seven six
   c seven point nought four four
   d nought point nought five
   e eighteen point three six
   f one point four nought seven

2 Write down these decimals in words.
   a 3.88
   b 6.03
   c 0.065
   d 11.07

3 Write down the value of the 7 in each of the following numbers.
   a 3.07
   b 73
   c 30.07
   d 2.74
   e 57.5
   f 0.007

> The first decimal place shows tenths, the second hundredths, and so on.

4 Write down the value of the 4 in each of the following numbers.
   a 6.04
   b 94
   c 72.45
   d 2.74
   e 5.004
   f 0.345

5 Look at the number 5.073. Which digit gives
   a the number of tenths
   b the number of thousandths?

6 Write > or < between each pair of numbers to make a true statement.
   a 3.57 ☐ 3.59
   b 25.64 ☐ 25.46
   c 4.88 ☐ 5
   d 6.74 ☐ 6.71
   e 85.37 ☐ 85.73
   f 6.33 ☐ 6

7 Write each set of numbers in order of size with the smallest number first.
   a 6.76, 4.83, 6.29
   b 9.07, 9.51, 9.18, 9.03

8 Write each set of numbers in order of size with the largest number first.
   a 12.6, 14.09, 12.55, 13.75
   b 7.555, 7.5, 7.05, 7.55

9 Write these numbers in order of size with the smallest number first.
   0.22, 2.2, 0.202, 0.022

10 Write these fractions as decimals.
   a $\frac{4}{10}$
   b $\frac{15}{100}$
   c $\frac{37}{100}$
   d $\frac{2}{10}$

11 Write these numbers in order of size with the largest number first.
   0.55, 0.055, 0.505, 0.555

5.2 Changing decimals to fractions

The positions of the figures after the decimal point tell you their value.
You can use this to write decimals as **fractions**.

Example 4

Write 0.15 as a fraction.

$$0.15 = \frac{1}{10} + \frac{5}{100} = \frac{15}{100} \overset{\div 5}{\underset{\div 5}{=}} \frac{3}{20}$$

Check: $\frac{3}{20} = 3 \div 20 = 0.15$

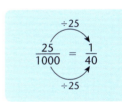
0.15 means $\frac{1}{10} + \frac{5}{100}$.
$\frac{1}{10} = \frac{10}{100}$

Example 5

Write 1.025 as a fraction.

$$1.025 = 1 + \frac{2}{100} + \frac{5}{1000}$$
$$= 1 + \frac{20}{1000} + \frac{5}{1000} = 1 + \frac{25}{1000}$$
$$= 1\frac{25}{1000}$$
$$= 1\frac{1}{40}$$

$$\frac{25}{1000} \overset{\div 25}{\underset{\div 25}{=}} \frac{1}{40}$$

Exam practice 5B

1 Write each decimal as a fraction in its lowest terms.
   a 0.2   b 0.5   c 0.06
   d 0.7   e 0.8   f 0.025

Remember that to give a fraction in its lowest terms means you have to cancel it as far as you can.

2 Write each decimal as a fraction in its simplest form.
   a 0.25   b 0.08   c 1.4
   d 0.125   e 2.06   f 5.005

3

0 .		

   2   0   5

   Copy the diagram, then write the numbers on the cards in the
   empty boxes so that the decimal number is
   a the largest possible   b the smallest possible.

4 0.85 of the families in Northgate Street own a car.
   What fraction of the families own a car?

5 In a wood, 0.68 of the trees lose their leaves in the winter.
   Find the fraction of the trees that lose their leaves.

6 George did a survey.
   He found that 0.48 of the cars that passed him had one passenger.
   Work out the fraction of the cars that had one passenger.

5.3 Addition and subtraction of decimals

Decimals can be added or subtracted in the same way as whole numbers. You add hundredths to hundredths, tenths to tenths, and so on.
When the numbers are simple you can do the calculation in your head.

Example 6

Work out $1.6 + 3 + 0.05$.

```
  1.60
  3.00
+ 0.05
------
  4.65
```

Write the numbers in a column with the decimal points under one another. Add 0s so that each number has the same number of decimal places.

Example 7

Calculate $3.5 - 1.06$.

```
  3.50
- 1.06
------
  2.44
```

Write the numbers in a column and write 3.5 as 3.50.

Exam practice 5C

1   Work in your head and just write down the value of:
    a $1.6 + 0.3$ **b** $2.3 + 0.05$ **c** $2.8 - 1$ **d** $1.6 - 0.22$
    e $0.24 + 1.7$ **f** $3 - 1.7$ **g** $0.79 - 0.38$ **h** $1.2 - 0.07$

2   Find the value of:
    a $1.77 + 3.9$ **b** $0.5 - 0.04$ **c** $2.44 - 1.74$
    d $8.04 - 3.27$ **e** $2.3 + 0.4 + 7.8$ **f** $7.5 + 1.44 + 3.06$
    g $6 - 3.8 + 1.5$ **h** $7.2 + 2.7 - 5.6$

3   After a walk in the rain, Ellen's coat weighed 2.45 kg.
    After it had dried out it weighed 0.3 kg less.
    What was the weight of the dry coat?

4   Sonia weighed 94.3 kg. She went on a diet.
    During the first three weeks she lost 2.35 kg.
    How much did she weigh then?

5   Zak weighed 105 kg. He went on a diet.
    He lost 9.5 kg in four months.
    How much did he weigh then?

6 This diagram shows a triangular field.
   What is the distance around the edge of the field?

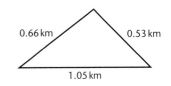

0.66 km   0.53 km

1.05 km

7 At the beginning of term Henry was 124.3 cm tall.
   By the end of term he was 126 cm tall.
   How much did he grow during the term?

8 The diagram shows a rectangle.
   How far is it around the edge of
   this rectangle?

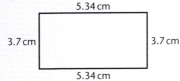

5.34 cm

3.7 cm   3.7 cm

5.34 cm

9 The bill for three meals was £9.
   The first meal cost £2.43 and the second meal cost £3.72.
   What was the cost of the third meal?

10 Rik bought 4 items. This is his bill.
    a Copy and complete the bill.
    b Rik paid with a £10 note.
       How much change should he get?

£1.30
£2.50
£0.95
£1.34

TO PAY _____

11 A stick is 28.3 centimetres long.
    Three pieces are cut.
    Their lengths are 8.3 cm, 2.8 cm and 7.7 cm.
    a What is the total length cut off?
    b What length is left?

12 A piece of wood is 20.2 mm thick.
    It is planed to make it smooth.
    The first planing takes 1.08 mm off the thickness.
    The second planing takes off another 0.34 mm.
    Work out the new thickness.

13 A length of metal passes through a set of rollers.
    The first pass reduces its thickness by 0.44 mm, the second pass
    by 0.33 mm and the third pass by 0.25 mm.
    The metal is now 8.2 mm thick.
    How thick was it to start with?

14 The distances a lorry travelled between deliveries were
    31.2 km, 27.5 km, 9.9 km and 16.3 km.
    This is the display at the last delivery point.

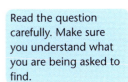

Read the question
carefully. Make sure
you understand what
you are being asked to
find.

80
60   100
40 0 0 3 7 2 3·7 120
20   140
     Km/h
   0   160

What did it show at the first delivery point?

5.4 Multiplying and dividing by 10, 100, 1000, ...

You can multiply and divide decimals by 10, 100, 1000, ... using a place value diagram.

1000s	100s	10s	1s	$\frac{1}{10}$ths	$\frac{1}{100}$ths
			1	2 • 7	2
		1	2	7 • 2	
	1	2	7	2 •	

12.72×10

12.72×100

When a number is multiplied by 10, 100, 1000, ... the figures move 1, 2, 3, ... places to the left on a place value diagram.

Example 8

Find **a** 0.45×100 **b** 0.45×1000.

a $0.45 \times 100 = 45$

b $0.45 \times 1000 = 450$

100s	10s	1s	$\frac{1}{10}$ths	$\frac{1}{100}$ths
		0 •	4	5
	4	5 •		
4	5	0 •		

$\times 100$
$\times 1000$

0 needed to show there are no units.

When a number is divided by 10, 100, 1000, ... the digits move 1, 2, 3, ... places to the right on a place value diagram.

Example 9

Find **a** $45 \div 10$ **b** $45 \div 1000$.

a $45 \div 10 = 4.5$

b $45 \div 1000 = 0.045$

10s	1s	$\frac{1}{10}$ths	$\frac{1}{100}$ths	$\frac{1}{1000}$ths
4	5 •			
	4 •	5		
	0 •	0	4	5

$\div 10$
$\div 1000$

0 needed to show there are no tenths.

Exam practice 5D

1 Work out:
  a 2.5×100 **b** 0.066×10 **c** 24.4×1000 **d** 0.1×100

2 Find:
  a $4.6 \div 10$ **b** $0.85 \div 1000$ **c** $12 \div 10$ **d** $9.6 \div 100$

3 Look at the number 25.73.
Calculate the value of the digit 7 multiplied by twice the value of the digit 5.

> The digit 7 is in the first decimal place so represents $\frac{7}{10}$.

4 Look at the number 12.08.
Subtract ten times the value of the digit 8 from twice the value of the digit 2.

5.5 Multiplication and division of decimals

You can multiply or divide a decimal by a whole number in the same way that you multiply and divide whole numbers.

Example 10

Work out: **a** 1.3×200 **b** $2.7 \div 200$.

a $1.3 \times 2 = 2.6$

$\qquad 2.6 \times 100 = 260$

$\qquad$ so $1.3 \times 200 = 260$

> To multiply by 200, first multiply by 2 then multiply your answer by 100.

b $\begin{array}{r} 1.35 \\ 2\overline{)2.70} \end{array}$

$1.35 \div 100 = 0.0135$

so $2.7 \div 200 = 0.0135$

> Divide by 2 then divide the answer by 100. Add zeros when you need to continue the division.

When you divide a whole number by another whole number, you can use decimals to continue the division.

To divide by a decimal, write the calculation as a fraction and find an equivalent fraction with a *whole number* on the bottom.

Example 11

Find: **a** the exact value of $135 \div 8$ **b** $2.8 \div 0.25$.

a $\begin{array}{r} 16.875 \\ 8\overline{)135.000} \end{array}$

$135 \div 8 = 16.875$

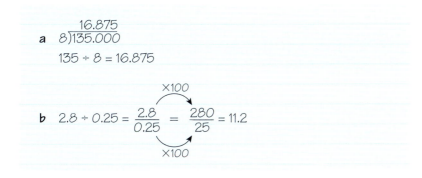

b $2.8 \div 0.25 = \dfrac{2.8}{0.25} = \dfrac{280}{25} = 11.2$

> $2.8 \div 0.25$ can be written as the fraction $\dfrac{2.8}{0.25}$. Change this to an equivalent fraction with denominator 25 by multiplying top and bottom by 100.

To multiply decimals you can use this rule:

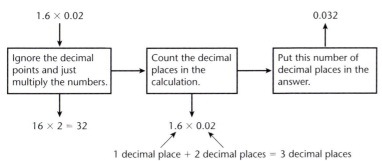

Example 12

Calculate 1.5×0.2.

$15 \times 2 = 30$

so $1.5 \times 0.2 = 0.30 = 0.3$

> 1.5×0.2 gives 1 decimal place + 1 decimal place = 2 decimal places so put a decimal point in 30 to give 2 decimal places.

Exam practice 5E

> To multiply by 300, multiply by 3 then multiply the answer by 100.

1 Find:
a 0.4×4
b 1.6×2
c 2.4×3
d 1.3×5
e 0.3×300
f 1.6×20
g 1.2×30
h 0.7×200

> To divide by 10, 100,.. move the digits one, two, ... places to the right.

2 Find:
a $1.5 \div 50$
b $14.4 \div 120$
c $25 \div 5000$
d $0.84 \div 40$

3 Work out:
a 1.6×0.3
b 0.12×0.2
c 0.22×0.04
d 1.05×0.002
e 1.2×0.1
f $(0.1)^2$
g 45×0.01
h $(0.01)^2$

4 Work out:
a $0.3 \div 0.1$
b $2 \div 0.2$
c $1.4 \div 0.07$
d $0.36 \div 1.2$
e $0.8 \div 0.04$
f $6.3 \div 0.21$
g $6 \div 0.15$
h $1.5 \div 0.75$

5 Find:
a half of 6.48 kg
b one-third of 39.36 cm
c one-quarter of 248 mm
d one-fifth of 18.9 m.

> To find $\frac{1}{3}$ of a quantity divide by 3.

6 Find the product of the value of the digit 5 and the value of the digit 9 in the number 0.95.

> **Product** means the result of multiplying numbers together.

7 Find:
a $9.9 \div 0.01$
b $26.7 \div 0.08$
c $10.5 \div 1.5$
d $1.08 \div 1.2$
e 2.56×1.2
f $3.6 \div 0.12$
g 4.5×0.03
h $0.056 \div 0.8$

8 Work out:
a $8.4 \div 21$
b 1.33×3.2
c $4.5 \div 1.5$
d 0.71×6.3
e 400×1.6
f $500 \div 0.25$
g 1200×0.001
h $14.7 \div 0.2$

9 Work out:
a $(0.5)^3$
b $(0.7)^3$
c $(0.4)^3$

> Remember:
> $(0.5)^3 = 0.5 \times 0.5 \times 0.5$

10 a Find the cost of 26 books at £6.55 each.
b Find the cost of 5.4 m of wood at 35p per metre.

11 A company employs 8 people.
Each person is paid £372.55 a week.
Work out the weekly wage bill for the 8 employees.

12 Wood costs 2.6p per cubic centimetre.
Find the cost of 85 cubic centimetres of wood.

13 16 coins are placed in a pile.
Each coin is 1.23 millimetres thick.
Work out the height of the pile.

14 A piece of wire is 12.55 metres long.
It is cut into 50 pieces of equal length.
How long is each piece?

15 One fence panel is 1.5 metres long.
How many of these panels are needed to make a fence 42 metres long?

16 £16.36 is divided equally between four boys.
How much does each boy receive?

17 A shelf is 247 centimetres long.
Books are to be stored on this shelf.
Each book is 2.6 centimetres thick.
How many of these books will fit on this shelf?

5.6 Bills

Most calculations with money use decimals.

Exam practice 5F

1 Lyn went shopping for a party.
Copy and complete her bill.

60 chocolate bars @ 44p each   £ [.   ]

45 cans of squash @ 42p each   £ [.   ]

[   ] packets of crisps @ 32p each   £ 7.04

Total   £ [.   ]

> You have to find the number of packets of crisps that cost a total of £7.04.

2 Tim is decorating his lounge.
The wallpaper costs £10.55 a roll.
The paint £8.33 a litre.
Paste costs £1.34 a packet.

a Copy and complete Tim's bill:

9 rolls of wall paper   £ [.   ]

2 litres of paint   £ [.   ]

[   ] packets of paste   £ 4.02

Total   £ [.   ]

b In a sale the same wallpaper is marked '$\frac{1}{3}$ off'.
How much would Tim have saved by waiting for the sale?

3   Mrs Brown goes shopping.
    Grapefruit cost 32p each.
    Oranges cost 17p each.
    Cherries cost £4.02 a kilogram.

www   a   Copy and complete her bill

        5 grapefruit   £ [.]
        8 oranges   £ [.]
        [   ] kg cherries   £ 2.01
                            Total   £ [.]

       b   She paid with a £10 note.
           How much change did she get?

4   George bought trainers, socks and T-shirts for his children.
    Trainers cost £32.50 a pair.
    T-shirts cost £7.45 each.
    Socks cost £1.55 a pair.

www   a   Copy and complete George's bill.

        3 pairs of trainers   £ [.]
        5 T-shirts   £ [.]
        [   ] pairs of socks   £ 7.75
                            Total   £ [.]

       b   Normal trainers cost £32.50 a pair.
           Discount trainers were sold at half-price.
           What is the total cost of 3 pairs of discount trainers?

5   In a supermarket:
    new potatoes cost £2.80 / kg
    carrots cost £1.46 / kg
    parsnips cost £1.58 / kg
    swedes cost £1.44 / kg.

www   a   Copy and complete this bill

        2 kg new potatoes   £ [.]
        $\frac{1}{2}$ kg carrots   £ [.]
        $1\frac{1}{2}$ kg parsnips   £ [.]
        [   ] kg swede   £ 0.72
                            Total   £ [.]

       b   At another shop the potatoes were half-price and carrots cost
           £1.36 / kg.
           Parsnips and swedes cost the same.
           Work out the bill for the same shopping at this shop.

6 Fred pays his newsagent's bill.
   a Copy and complete his bill.

   12 Daily Record at 45p each   £ [.   ]
   2 Sunday papers at £1.45 each   £ [.   ]
   3 magazines at £3.55 each   £ [.   ]
   [   ] choc ices at 99p each   £ 2.97
                                 Total   £ [.   ]

   b The price of the Daily Star went up by 5p a copy.
      The price of the Sunday paper went up by 10p a copy.
      The prices of the magazines and the ice cream stayed the same.
      Work out Fred's bill using the new prices.

5.7 Changing fractions to decimals

The fraction $\frac{3}{8}$ means $3 \div 8$ so $\frac{3}{8}$ can be written as a decimal by working out $3 \div 8$:

$$\begin{array}{r} 0.375 \\ 8)\overline{3.00} \end{array}$$

So $\frac{3}{8} = 0.375$.

To change a fraction to a decimal, divide the numerator by the denominator.

Recurring decimals

If you try to write $\frac{1}{3}$ as a decimal, you get 0.33333 … and so on for ever.

0.3333 . . . is called a **recurring decimal**. It is written $0.\dot{3}$.

Any decimal with a recurring digit or pattern of digits is called a recurring decimal.

When you convert a fraction to a decimal, you will either get an exact decimal or a recurring decimal.

The digit 3 recurs.

A dot over a digit means that it recurs.
So $0.\dot{5}$ means 0.5555555...

$\frac{1}{13} = 0.076923076923076...$ where 076923 recurs.
You write this as $0.\dot{0}7692\dot{3}$.
The first overhead dot to the second overhead dot shows the pattern of digits that repeat.

To convert a fraction to a decimal divide the top by the bottom. Carry on until it stops or recurs.

Exam practice 5G

1 Write each fraction as a decimal.
   a $\frac{1}{5}$ b $\frac{1}{8}$ c $\frac{3}{4}$ d $\frac{3}{5}$

2 Write each fraction as a decimal.
   a $\frac{4}{5}$ b $\frac{3}{8}$ c $\frac{3}{20}$
   d $\frac{1}{4}$ e $\frac{7}{8}$ f $\frac{6}{25}$

3 a Write $\frac{9}{20}$ as a decimal.
   b Which is larger: $\frac{9}{20}$ or 0.47?

4 a Write $\frac{3}{4}$ as a decimal.
   b Which is larger: $\frac{3}{4}$ or 0.72?

5 Arrange these in order of size with the smallest first.
   $\frac{1}{4}$, 0.35, $\frac{1}{8}$, 0.2

6 Write > or < between each pair of numbers to make a true statement.

   a $\frac{2}{5}$ □ 0.3   b $\frac{4}{5}$ □ 0.78   c $\frac{7}{8}$ □ 0.79

   d $\frac{5}{8}$ □ 0.6   e 0.66 □ $\frac{2}{3}$ f 1.8 □ $\frac{11}{6}$

> You need to calculate three decimal places for part **e**.

7 Arrange these numbers in order with the largest first.
   0.05, $\frac{3}{16}$, 0.105, $\frac{2}{13}$, $\frac{6}{25}$

8 Find the reciprocal of each of the following.
   Give your answer as a decimal.

   a 0.2   b $\frac{4}{5}$ c 1.6   d $2\frac{1}{2}$ e 1.5

> The **reciprocal** of 0.2 is 1 divided by 0.2. The reciprocal of $\frac{4}{5}$ is $\frac{5}{4}$.

9 Express each fraction as a recurring decimal.

   a $\frac{2}{3}$ b $\frac{1}{7}$ c $\frac{1}{6}$ d $\frac{4}{15}$

   e $\frac{2}{9}$ f $\frac{1}{11}$ g $\frac{1}{12}$ h $\frac{5}{13}$

10 Write down the recurring decimal $0.\dot{3}1\dot{6}$ to 9 decimal places.

11 The value of $\frac{6}{13}$ as a recurring decimal is $0.\dot{4}6153\dot{8}$.
    Write $\frac{6}{13}$ as a decimal to 12 decimal places.

12 Use dot notation to write these fractions as decimals.

   a $\frac{4}{30}$ b $\frac{4}{300}$ c $\frac{4}{3000}$

5.8 Standard form

It is difficult to compare the size of these two numbers because one is in billions and the other is in millions.

Very large numbers or very small numbers are easier to compare when they are written using the same notation.

The notation that is used in science is called **standard form**.

A number written in standard form is **a number between 1 and 10 multiplied by a power of 10**.

CONSUMER CREDIT STANDS AT £52.6 BILLION

£50 MILLION BANK HEIST

> The numbers 1.3×10^2 and 3.72×10^{-3} are in standard form.
> The numbers 13×10^5 and 0.26×10^{-3} are not in standard form because the first number is not between 1 and 10.

Example 13

Write 5.976×10^{24} in full.

$5.976 \times 10^{24} = 5\,976\,000\,000\,000\,000\,000\,000\,000\,000.$

> 5.976×10^{24} means 5.976 multiplied by ten 24 times. This means you have to move the digits 24 places to the left. Fill in the gaps with zeros.

Exam practice 5H

1 Write the following in full.
   - a 5.5×10^3
   - b 3.16×10^5
   - c 4.155×10^6
   - d 5.778×10^2
   - e 1.3×10^4
   - f 9.15×10^3
   - g 8.022×10^4
   - h 2.004×10^8
   - i 7.4×10^6

> To multiply by 10, 100, 1000, … **move the figures** 1, 2, 3,… places to the **left** and fill in any gaps with zeros. For example
> $3.07 \times 10^5 = 3.07 \times 100\,000$
> $= 3.07000 \times 100\,000 = 307\,000$

2 A noise level of 10^{12} decibels is painful.
   Write this noise level in full.

3 The Earth is 1.5×10^8 kilometres from the sun.
   Write this distance in full.

4 When the number 1.225×10^{12} is written as an ordinary number, how many zeros are there after the 5?

5 a Without using a calculator, find $1\,200\,000 \times 40\,000$.
   b Now use your calculator to find $1\,200\,000 \times 40\,000$.
      Write down exactly what is showing on the display of your calculator. What do you think it means?

6 Use your calculator to find $101\,000 \times 20\,000\,000$.
   Write your answer in full.

5.9 Using π in exact calculations

π is used in calculations involving the area and circumference of circles. You can do these calculations without a calculator, but to give an exact answer you must leave π in your answer.

Example 14

Find the exact value of $2 \times 1.6 \times \pi$.

$2 \times 1.6 \times \pi = 3.2 \times \pi$

> Multiply 2 by 1.6 but leave π as it is.

Exam practice 5I

1 Find the exact value of:
   - a $5 \times 0.8 \times \pi$
   - b $\pi \times (1.2)^2$
   - c $2 \times 1.32 \times \pi$
   - d $\pi \times (0.9)^2$
   - e $10 \times \pi \times 5.96$
   - f $(1.5)^2 \times \pi$

2 Xang said 'π is about 3 so $4 \times \pi = 12$'.
   Explain why Xang's answer is not exact.

Summary of key points

- The decimal point divides the units from the tenths.
- You can add and subtract decimals by writing them in columns; make sure that the decimal points are in line.
- You can multiply decimals by 10, 100, ... by moving the digits 1, 2, ... places to the left, for example $3.45 \times 10 = 34.5$ and $12.8 \times 100 = 1280$.
- You can divide decimals by 10, 100, ... by moving the digits 1, 2 ,... places to the right, for example $75.3 \div 10 = 7.53$ and $5.03 \div 100 = 0.0503$.
- You can divide a decimal by a whole number using the same method you use for whole numbers; keep the decimal point in the answer above the decimal point in the number you are dividing into.
- A fraction can be changed to a decimal by dividing the numerator by the denominator.
- A decimal can be changed to a fraction by writing the numbers after the point as tenths, hundredths, ... and simplifying.
- You should learn that $\frac{1}{2} = 0.5$, $\frac{1}{4} = 0.25$, $\frac{3}{4} = 0.75$ and $\frac{1}{8} = 0.125$.
- When you multiply decimals, ignore the decimal points and multiply the numbers; then the number of decimal places in the answer is the sum of the decimal places in the numbers that are multiplied together.
- To divide by a decimal, multiply the top and bottom by the same number so that the denominator becomes a whole number.
- Sizes of numbers can be compared by converting fractions to decimals.
- Some fractions give exact decimals and some give recurring decimals.
- 5.6×10^2 means 5.6×100.

Most students who get GRADE E or above can:
- add and subtract decimals
- multiply by decimals.

Most students who get GRADE C can also:
- divide by a decimal.

Glossary

Decimal place	a digit to the right of the decimal point
Denominator	the bottom number of a fraction
Fraction	part of the whole
Numerator	the top number of a fraction
Place value	the position of a digit that indicates its value
Product	the result of multiplying two quantities together
Recurring decimal	a decimal with a recurring pattern of digits, e.g. 0.12121212... written as $0.\dot{1}\dot{2}$
Standard form	any number written as a number between 1 and 10 multiplied by a power of ten

6 Approximations and estimation

<table>
<tr><td>

This chapter will show you:
- ✓ the meaning of a significant figure
- ✓ how to round a number to a given place value
- ✓ how to round to a given number of decimal places
- ✓ how to estimate the value of a calculation

</td><td>

Before you start you need to know:
- ✓ about place value
- ✓ the meaning of squares and cubes of numbers
- ✓ how to add, subtract, multiply and divide decimals
- ✓ the meaning of square root
- ✓ what a square number is

</td></tr>
</table>

6.1 Rounding numbers to the nearest 1, 10, 100, …

The number 167 means 1 hundred, 6 tens and 7 units.
To give 167 to the nearest ten means you have to decide if 167 is nearer
1 hundred and 6 tens or 1 hundred and 7 tens.
The arrow on the number line shows the number 167.

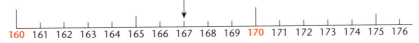

160 161 162 163 164 165 166 167 168 169 170 171 172 173 174 175 176

You can see that 167 is nearer 170 than 160.
So 167 = 170 to the nearest ten.

Giving a number to the nearest ten is called **rounding** the number
to the nearest ten or writing the number correct to the nearest ten.

Numbers can be rounded to the nearest 100, the nearest 1000, the
nearest unit, and so on.

Rounding is approximating.

Example 1

a Write 3625 correct to the nearest hundred.
b Round 46.5 to the nearest unit.

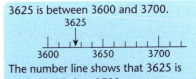

3625 is between 3600 and 3700.

The number line shows that 3625 is nearer 3600 than 3700.

   a   3625 = 3600 correct to the nearest hundred.

   b   46.5 = 47 to the nearest unit.

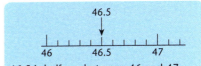

46.5 is halfway between 46 and 47.
When this happens, always round up.

You do not have to draw a number line each time you want to round a number.

You can use these rules to round numbers without using a number line:

- Find the figure you want to round to and draw a line after it.
- Look at the figure after the line.
  If it is 5 or more, round up.
  If it is less than 5, round down.

Example 2

Give

a 26 to the nearest 10

b 152 to the nearest 10

c 2635 to the nearest 10

d 96 to the nearest 10.

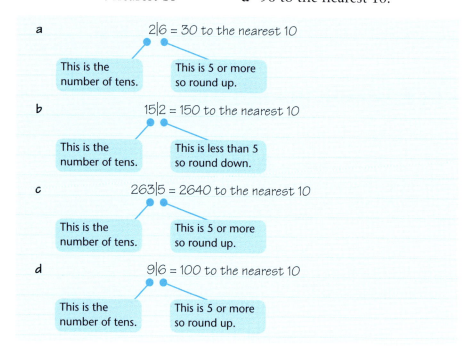

Start by drawing a line after the 2.

Example 3

a Give 8423 correct to the nearest 100.

b Give 4.8 to the nearest unit.

c Give 5673 to the nearest 1000.

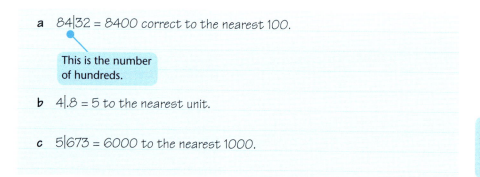

Draw a line after the thousands.
5|673 = 6000 to the nearest 1000.

Exam practice 6A

1 Round each number to the nearest ten.
   a 228   b 73   c 151
   d 632   e 37   f 155

2 Round each number to the nearest hundred.
   a 256   b 1221   c 450
   d 3780   e 1965   f 7639

3 Round each number to the nearest unit.
   a 8.2   b 3.8   c 5.5
   d 1.8   e 18.6   f 0.8

4 Round each number to the nearest thousand.
   a 7450   b 3025   c 6952
   d 15 701   e 28 040   f 1860

5 a Write 6782 to the nearest 10.
   b Write 6782 to the nearest 100.
   c Write 6782 to the nearest 1000.
   d Write 95 678 to the nearest 1000.
   e Write 95 678 to the nearest 100.
   f Write 95 678 to the nearest 10.

6 a Write 159.5 to the nearest 100.
   b Write 159.5 to the nearest 10.
   c Write 159.5 to the nearest unit.
   d Write 2177 to the nearest ten.
   e Write 2578 to the nearest hundred.
   f Write 439 to the nearest ten.

7 54 263 people went to a football match.
   Give this number to the nearest 1000.

8 £25 886 was collected for charity.
   Give this amount to the nearest £100.

9 The 2001 census showed that the population of the UK was
   58 789 194.
   How many people is this to the nearest million?

6.2 Rounding to a given number of decimal places

Giving a number to the nearest tenth, hundredth and so on, is done in the same way as giving a number to the nearest ten, hundred, and so on.

To give a number correct to a number of decimal places, draw a line after that decimal place and look at the figure after the line.
If it is less than 5, round down.
If it is 5 or more, round up.

Giving a number to the nearest tenth is called **rounding to one decimal place**. Giving a number to the nearest hundredth is called **rounding to two decimal places**.

Example 4

Give 1.38 correct to 1 decimal place.

1.3|8 = 1.4 correct to 1 decimal place.

Draw a line after the first decimal place.

8 is more than 5, so round up.

Example 5

Give **a** 0.7643 correct to 2 decimal places
      b 0.02973 correct to 3 decimal places.

a   0.76|43 = 0.76 correct to 2 d.p.

Draw a line after the 2nd decimal place. The next figure is 4 so round down.

This is the second decimal place.

d.p. is short for decimal place.

Draw a line after the 3rd decimal place. The next figure is 7 so round up.
The last 0 must be included to show that you have rounded to 3 d.p.

b   0.029|73 = 0.030 correct to 3 d.p.

Exam practice 6B

1 Write these numbers correct to 1 decimal place.
   a 2.48   **b** 12.62   **c** 124.77
   d 24.245   **e** 3.1689   **f** 0.0577

2 Write these numbers correct to 2 decimal places.
   a 6.573   **b** 0.347   **c** 12.655
   d 0.0329   **e** 4.8021   **f** 0.8005

3 Write these numbers correct to 3 decimal places.
   a 1.5568   **b** 0.4813   **c** 2.3557
   d 0.0153   **e** 0.1041   **f** 0.00407

4 Write
   a 0.6942 correct to 2 d.p.   **b** 40.378 correct to 2 d.p.
   c 28.75 correct to 1 d.p.   **d** 13.479 correct to 1 d.p.
   e 0.9999 correct to 2 d.p.   **f** 77.99841 correct to 3 d.p.

d.p. is short for decimal place.

5 Write
   a 2.2525 correct to 3 d.p.
   b 1.002793 correct to 4 d.p.
   c 0.0507702 correct to 2. d.p.
   d 298.2 to the nearest ten
   e 39.87 to the nearest unit
   f 139.78 correct to 1 d.p.

6 Write
   a 905.8 to the nearest ten
   b 0.479 correct to 2 d.p.
   c 375.57 to the nearest hundred.

7 Helen worked out the diameter of a circle on her calculator.
   Her display showed

   14.897667

   Give this number correct to 1 decimal place.

8 It takes light 1.26666 seconds to travel from the Moon to the
   Earth.
   Give this time correct to 3 decimal places.

6.3 Rounding answers

You are often asked to give answers to a certain degree of accuracy.
You should always say if you have rounded your answer.

Example 6

A bag of 30 sweets costs £1.40. Find the cost of one sweet to the
nearest penny.

Cost of 1 sweet is £1.40 ÷ 30 = £0.04|6...

| You usually give money answers to the nearest penny.

Draw a line after the number of pence.

= £0.05 to the nearest penny.

Example 7

A pile of 12 identical books stands 34.6 cm high.
How thick, correct to 1 decimal place, is 1 book?

Thickness of 1 book = 34.6 cm ÷ 12
               = 2.8|8... cm
               = 2.9 cm correct to 1 d.p.

Example 8

Write $\frac{5}{12}$ as a fraction correct to 2 decimal places.

$\frac{5}{12}$ = 0.41|6... = 0.42 correct to 2 d.p.

You need to find 5 ÷ 12.
Using a calculator, the display shows 0.41666666...
Write down the first 3 decimal places.

Exam practice 6C

1 A stack of 6 concrete posts weighs 275 kg.
Find the weight of 1 post.

> Give your answer correct to the nearest kilogram.

2 A stack of 500 sheets of paper is 4.6 cm high.
Work out the thickness of 1 sheet of paper.
Give your answer correct to 3 decimal places.

3 a Tim buys a pack of 6 cans of cola. He pays £2.54.
How much does one can cost?

> Give your answers correct to the nearest penny.

   b A pack of 12 apples costs £1.56. Find the cost of 1 apple.

4 a A van costs £45 to hire for 3 days. What is the cost per day?

   b Wood costs 2.5p per cubic centimetre.
Find the cost of 85.4 cubic centimetres of wood.

5 Write as a decimal: **a** $\frac{1}{3}$ **b** $\frac{1}{7}$.

> Give your answers correct to 2 d.p.

6 Write each fraction as a decimal correct to 2 decimal places.

   a $\frac{2}{3}$ **b** $\frac{1}{9}$ **c** $\frac{1}{6}$ **d** $\frac{4}{15}$

   e $\frac{2}{9}$ **f** $\frac{1}{11}$ **g** $\frac{1}{12}$ **h** $\frac{5}{13}$

7 Find the value of **a** 4.4×12.7 **b** $3.8 \div 1.6$.

> Give your answers correct to 1 d.p.

8 Find the value, correct to 3 decimal places, of

   a $0.7 \div 3$ **b** $0.23 \div 9$ **c** $0.013 \div 7$ **d** $\frac{14}{9}$.

> Remember that $\frac{14}{9}$ means $14 \div 9$.

9 Find the value of $5.5 - 2.3 \div 1.4$ correct to 1 decimal place.

> Remember to do the division first.

10 Write $>$ or $<$ between each pair of numbers to make a true statement.

   a $\frac{1}{3} \ \square \ 0.3$ **b** $\frac{4}{5} \ \square \ 0.75$ **c** $\frac{7}{8} \ \square \ 0.85$

   d $0.7 \ \square \ \frac{2}{3}$ **e** $0.57 \ \square \ \frac{4}{7}$ **f** $1.75 \ \square \ \frac{11}{6}$

> Change the fractions to decimals correct to 3 d.p.

11 a Express each of the following as a decimal correct to 3 decimal places.

       $\frac{5}{3}, \ 1\frac{2}{7}, \ \frac{15}{11}$

   b Arrange these numbers in order with the smallest first.

       $1.57, \ \frac{5}{3}, \ 1.49, \ 1\frac{2}{7}, \ \frac{15}{11}$

> You need to find 4 decimal places to give an answer correct to 3 d.p.

12 Arrange these numbers in order with the largest first.

   $0.04, \ \frac{3}{8}, \ 0.205, \ \frac{5}{16}, \ \frac{7}{25}$

13 Work out the value of each of the following correct to 1 decimal place.

   a $\dfrac{1.7 \times 3.5}{2.6}$ **b** $\dfrac{3.7}{1.2 \times 2.3}$

   c $\dfrac{1.8 - 0.9}{3.5}$ **d** $\dfrac{5.4 - 0.6}{2.7}$

> You can do each of these in one step on your calculator. Use brackets round the part that needs to be done first.
> For **a** press
> (1 . 7 × 3 . 5) ÷ 2 . 6 =

14 a   16 coins are placed in a pile. The pile is 21.3 millimetres high.
How thick is each coin? Give your answer in millimetres
correct to 1 d.p.

b   A piece of wire is 12.8 metres long. It is cut into 50 pieces of
equal length. How long is each piece? Give your answer in
centimetres correct to 1 d.p.

A01

15   Lara was told that it would take her 1 hour, to the nearest hour,
to drive from Bristol to Cardiff. What is the shortest time she
could expect to take?
Give a reason for your answer.

A01

16   Bella said she weighed 51 kg to the nearest kilogram.
Javid said 'You could weigh as much as 52 kg'.
Is Javid correct? Explain your answer.

6.4 Significant figures

Peter measured the thickness of a book.
He wrote down 0.0205 metres.

Jane measured the same book.
She wrote down 20.5 millimetres.

The measurements are the same but they look different.
This is because the units of measurement are different.

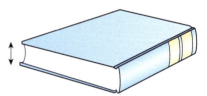

2.05 cm is also the
same length.

The important, or significant, figures are the 2 and the figures that
follow it.
The most important figure is 2 because it is the highest place
value figure in both numbers that is not zero. It is called the **first
significant figure**.

Example 9

Write down the first significant figure in each number and give its
value.
a   450   **b**   0.72

a   4. This figure is in the hundreds position, so its value is 400.

b   7. This figure is in the first decimal place so its value is 7 tenths.

**To give a number correct to one significant figure, draw
a line after that significant figure and look at the figure
after the line.
If it is less than 5, round down.
If it is 5 or more, round up.**

Example 10

Write   **a**   0.0635 correct to 1 significant figure
        b   7.773 correct to 1 significant figure
        c   34.507 correct to 1 significant figure.

a   *0.06|35 = 0.06 correct to 1 s.f.*

s.f. is short for significant figures.

b   *7|.773 = 8 correct to 1 s.f.*

c   *3|4.507 = 30 correct to 1 s.f.*

Exam practice 6D

1   Write down the first significant figure in each number and give its value.

a 43	b 9.2	c 255
d 0.82	e 0.065	f 24.88
g 20.03	h 502.2	i 0.047

2   Round each number to one significant figure.

a 25.8	b 0.57	c 7967
d 580	e 1.8	f 5.2
g 0.078	h 45.3	i 0.57

3   Give these numbers correct to 1 significant figure.

a 2693	b 37 251	c 67 600
d 72 505	e 9943	f 586
g 888	h 999	i 64.88
j 0.07643	k 0.006438	l 354.77
m 4.874	n 0.3762	p 10.555
q 0.004 6748	r 0.099	s 5.584

4   Find, without using a calculator, correct to 1 s.f.

a $10 \div 6$	b $75 \div 9$	c $0.44 \div 7$	d $147 \div 8$

6.5 Estimating answers to calculations

Estimating the value of a calculation gives you an **approximate** answer.
You can get an **estimate** by rounding each number to one significant figure.

Example 11

Estimate the value of:

a 28.8 ÷ 5.2   **b** 24.2 ÷ 2.87   **c** 6.871 ÷ 0.46

a 2|8.6 ÷ 5|.2 ≃ 30 ÷ 5 = 6

> Round each number to one significant figure then use the rounded numbers in the calculation.

This symbol means 'is approximately equal to'. Other symbols used are ≈ and ~.

b 2|4.2 ÷ 2|.87 ≃ 20 ÷ 3 ≃ 6.6 = 7 to 1 s.f.

> Round your estimate to 1 s.f.

c 6|.871 ÷ 0.4|6 ≃ 7 ÷ 0.5 = 70 ÷ 5 = 14 = 10 to 1 s.f.

Exam practice 6E

> In questions 1 to 6 start by rounding each number to one significant figure.

1 Work out an approximate value for:
   a 5.6×12.1 b $590 \div 9.1$ c 31×8.3
   d $97 \div 4.8$ e $870 \div 24$ f $5.8 \div 3.2$

2 Find an approximate value of:
   a 570×21 b $294 + 149$ c $2156 + 3907$
   d 0.37×0.14 e 0.43×0.27 f 4.49×0.75

3 Find an estimate for:
   a $(4.9)^2$ b $(0.037)^2$ c $(0.29)^3$ d $\sqrt{98}$

   > Remember that $(4.9)^2$ means 4.9×4.9 and that $(0.294)^3$ means $0.294 \times 0.294 \times 0.294$.

4 Write down an estimate for:
   a $\dfrac{1.97}{2.95}$ b $\dfrac{5.39}{0.045}$ c $\dfrac{0.432}{0.208}$ d $\sqrt{396}$

   > Remember that $\dfrac{1.97}{2.95}$ means $1.97 \div 2.95$.

5 Find an approximate value of:
   a $0.37 \div 0.14$ b $1.39 \div 0.045$
   c $\dfrac{0.027}{0.52}$ d $\dfrac{174}{0.37}$

6 Fayed bought 310 boxes of paper for his office.
   Each box cost £9.56.
   Estimate the total cost.

7 The answers given to these calculations are all wrong.
   Decide whether each answer is too big or too small.
   a $278 \div 37 = 751$ b $72 + 85 = 83$ c $5.62 \times 1.15 = 4.16$
   d $\dfrac{2 + 7}{911} = 8.2$ e $\dfrac{2.66}{5.7 - 2.3} = 7.6$ f $(1.9)^2 = 5.2$

8 Use estimates to choose the correct answer for each calculation.
   a 2.09×15.2:   **A** 3.1768   **B** 31.768   **C** 45.72   **D** 0.0663
   b $(2.09)^2$:   **A** 0.43681   **B** 437   **C** 25.9   **D** 4.37
   c $\dfrac{25.4}{2.8}$:   **A** 9.07   **B** 0.907   **C** 102.5   **D** 0.0102
   d $25 \times 42 \times 34$: **A** 357   **B** 35700   **C** 35.7   **D** 357 000

9 Find an approximate value of:

a $\dfrac{3.8 \times 5.2}{2.1}$ b $\dfrac{0.63 \times 2.6}{5.4}$ c $\dfrac{21 \times 4.2}{7.9}$

d $\dfrac{6.3 \times 0.185}{0.62}$ e $\dfrac{89 \times 0.079}{5.9}$ f $\dfrac{97 + 9.8}{41}$

10 Estimate the value of:

a $\dfrac{8.7}{5.7 + 5.9}$ b $\dfrac{0.52}{6.4 + 0.73}$ c $\dfrac{29 \times 315}{62 - 44}$

d $\dfrac{85 - 57}{5.8 \times 100}$ e $\dfrac{2.4^2}{2.7 + 3.9}$ f $\dfrac{4937 + 5216}{(2.8 + 5.9) \times 100}$

11 Estimate the value of:

a $\dfrac{2.8 \times 5.8}{8.9 - 3.7}$ b $\dfrac{86.66}{42.43 - 21.59}$ c $\dfrac{(7.2 + 3.2)^2}{1.9}$

d $\sqrt{38.4 + 7.2}$ e $\sqrt{\dfrac{0.357}{0.013}}$ f $21\sqrt{4.86 \times 5.13}$

> Estimate the number under the **square root** then round it to the nearest **square number**.

12 The number of rabbits on Ase Island is approximately 27 000.
The numbers are expected to increase to 27 000 × 1.8 next year.
a Estimate the number of rabbits next year.
b In three years' time the number is expected to be
27 000 × 1.8³.
Estimate the number of rabbits in three years' time.
How accurate do you think your estimate is?

A01 13 Tony used his calculator to find 2.57 + 8.36 × 0.19.
He wrote down the answer as 2.08 correct to 2 decimal places.
Explain how you know that Tony's answer is wrong.

6.6 Using a calculator

When you use your calculator, you will often find that there are
many more figures in the display than you need. You do not have to
write all these figures down. Write down one more figure than you
need for your answer.

Example 12

Use your calculator to find $\dfrac{1.75 + 0.924}{2.477}$ correct to 2 decimal places.

$\dfrac{1.75 + 0.924}{2.477} = 1.07|9\ldots =$

1.08 correct to 2 d.p.

> Use brackets around 1.75 + 0.924 so that the calculator does this first:
> ⟮ 1 . 7 5 + 0 . 9 2 4 ⟯ ÷ 2 . 4 7 7 =
> The display shows 1.079531692.
> To give this correct to 2 decimal places, write down the first 3 decimal places.

If you do your working in more than one step on your calculator,
use the memory to store answers you will need again. If you have an
answer key, you can enter the answer to the last calculation.

Example 13

Find $\dfrac{5.88}{1.22 \times 12.9}$ correct to 3 decimal places.

$\dfrac{5.88}{1.22 \times 12.9} = 0.373|6\ldots$

$= 0.374$ correct to

3 decimal places.

Work out 1.22×12.9 first: 15.738.

Then enter ⑤ . ⑧ ⑧ ÷ ANS = : $0.3736\ldots$

You could also use brackets to do the calculation in one step: press

⑤ . ⑧ ⑧ ÷ ((① . ② ② × ① ② . ⑨)) = .

The instructions work on a DAL calculator. If you have an older calculator, use the manual.

Exam practice 6F

1   Estimate each calculation, then use your calculator to find the answer correct to 2 d.p.

   a 63×2.752 b $40.3 \div 2.7$ c $400 \div 35$
   d $5703 \div 154$ e $34.2 \div 30.7$ f $57 \div 2.51$
   g $0.27 \div 0.52$ h 36.8×4.15 i $72 \div 6.75$

2   Find, correct to 2 d.p., the value of:

   a $0.366 - 0.37 \times 0.52$
   b $0.0526 \times 0.372 + 0.027$
   c $6.924 + 1.56 \div 0.00793$

Remember that you do multiplication and division before addition and subtraction. First make an estimate. This will tell you whether your calculator answer is likely to be correct.

3   Find the value of:

   a $0.82 - 0.34 \times 0.58$ b $24.78 \times 0.07 + 8.25$
   c $0.83 \times 0.61 - 0.27$ d $5.78 - 0.73 \times 2.24$

Give your answers correct to 2 decimal places.

4   Find the value of:

   a $32.03 \times (17.09 - 16.9)$ b $54.6 \times (22.05 - 8.17)$
   c $6.04 \div (1.958 - 0.872)$ d $0.51 \div (0.45 + 0.327)$

Give your answers correct to 2 decimal places.

5   Find, correct to 3 d.p., the value of:

   a $\dfrac{0.016}{1.62 - 0.897}$ b $\dfrac{0.034 + 1.3667}{1.3142}$

   c $\dfrac{57.2}{1.113 \times 5.906}$ d $\dfrac{24.6^2}{297}$

To find 24.6^2 on a calculator, press ② ④ . ⑥ x^2 = .

6   Calculate, correct to 2 d.p., the value of:

   a $\dfrac{2.5}{1.8 \times 0.77}$ b $\dfrac{4.55 + 3.32}{2.79}$

   c $\dfrac{0.51 + 1.7}{0.095}$ d $\dfrac{1.3^3}{2.6}$

To find 1.3^3 on a calculator, press ① . ③ x^y ③ = .

7   Find, correct to 2 decimal places,

   a $\sqrt{12}$ b $\sqrt{54}$ c $\sqrt{120}$ d $\sqrt{4.966}$
   e $\sqrt{0.8845}$ f $\sqrt{23.62}$ g $\sqrt{0.0092}$ h $\sqrt{0.000395}$.

To find a square root, press √ , then enter the number and press = .

8 Plant cells are grown in a laboratory.
   When conditions are perfect, the number of cells after 3 hours is given by $25 \times (1.5)^3$.
   Work out the number of cells after 3 hours.

9 Andy said that he could get 250 cups of coffee from a 375 g jar of instant coffee powder using 0.8 grams of powder per cup.
   A01
   a Explain how you know that he is wrong.
   b Find the number of cups that Andy can get.

10 The volume of a packing carton is $(5.25 \times 3.67 \times 1.99)$ m³.
   Find the volume.

11 a Round each number in $\dfrac{1.27}{1.23 - 0.97}$ to 1 significant figure.
   A01
   b Explain why you cannot use these rounded numbers to find an estimate for $\dfrac{1.27}{1.23 - 0.97}$.

12 To calculate $\dfrac{2758}{12.5 \times 16.2}$ Greg pressed these keys:

   $\boxed{2}\,\boxed{7}\,\boxed{5}\,\boxed{8}\,\boxed{\div}\,\boxed{1}\,\boxed{2}\,\boxed{.}\,\boxed{5}\,\boxed{\times}\,\boxed{1}\,\boxed{6}\,\boxed{.}\,\boxed{2}\,\boxed{=}$

   A01
   a This gives the wrong answer.
      Explain why.
   b Work out the correct answer rounded to the nearest whole number.

13 Goran worked out $\sqrt{5.8} + 2.9$ by pressing these keys.

   $\boxed{\sqrt{}}\,\boxed{5}\,\boxed{.}\,\boxed{8}\,\boxed{+}\,\boxed{2}\,\boxed{.}\,\boxed{9}\,\boxed{=}$

   A01
   a Explain why Goran got the wrong answer.
   b Work out the right answer, correct to 1 decimal place.

14 Find, correct to 2 d.p.,
   a $\sqrt{(6.45^2 - 9.46)}$ b $\sqrt{(57.34 - 5.77^2)}$
   c $\sqrt{(3.142 \times 6.447 - 16.22)}$.

Summary of key points

- To round a number draw a line after the place value you want to round to, then look at the next figure after the line. If it is 5 or more, round up. If it is less than 5 round down.
- The first significant figure in a number is the first figure from the left-hand end that is not zero.
- You can find an approximate value of a calculation by rounding each number to one significant figure.

Most students who get GRADE E or above can:
- find the value of $3.2^2 + \sqrt{4.4}$ correct to a number of decimal places.

Most students who get GRADE C can also:
- estimate the value of $\sqrt{(4.96^2 - 7.546)}$.

Glossary

Approximation	a rough value
Decimal place	the position of a figure after the decimal point
Estimate	a rough value
Rounding	giving a number to a certain degree of accuracy
Significant figure	the first significant figure of a number is the first digit that is not zero
Square number	a number whose square root is a whole number
Square root	a number which when multiplied by itself gives the starting number

7 Measures 1

This chapter will show you:

✓ the common metric units used for measuring length, weight and capacity
✓ the common Imperial units used for measuring length, weight and capacity
✓ the relationships between different units for measuring a quantity
✓ how to convert between different units
✓ how to work out value for money
✓ the range in which a rounded value lies

Before you start you need to know:

✓ how to multiply and divide by 10, 100, 1000
✓ how to add, subtract, multiply and divide with decimals
✓ how to find one quantity as a fraction of another
✓ how to find a fraction of a quantity

7.1 Metric units of length

Metric units of length in everyday use are
the kilometre (km), the metre (m),
the centimetre (cm) and the millimetre (mm).

> The letters in the brackets show the short way of writing these units.

The relationships between these units are
1 kilometre = 1000 metres,
1 metre = 100 centimetres = 1000 millimetres,
1 centimetre = 10 millimetres.

> **Did you know**
>
> that there are two systems of units that are used for length, weight and capacity? These are **metric units** and **Imperial units**.
>
> Metric units are the main units of measurement used in the United Kingdom.

You can use these relationships to convert the measurement of a length from one unit to another.

To convert to a smaller unit, you multiply. ●────

> There are more of the smaller units in a given length.

To convert to a larger unit, you divide. ●────

> There are fewer of the larger units in a given length.

This diagram shows these relationships and how to use them.

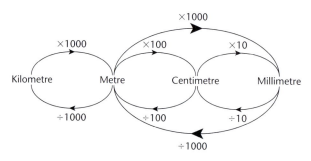

Example 1

a Convert 10.5 cm into millimetres.
b Convert 10.5 cm into metres.

a 10.5 cm = 10.5 × 10 mm
              = 105 mm

There are 10 millimetres in 1 cm so you need to multiply by 10.

b 10.5 cm = 10.5 ÷ 100 m
              = 0.105 m

There are 100 centimetres in 1 metre so you need to divide by 100.

Exam practice 7A

1 Write down the length of each strip of wood.

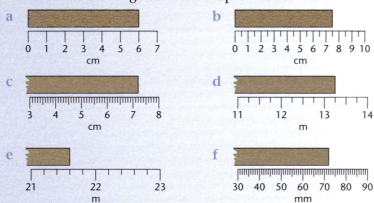

a   0 1 2 3 4 5 6 7
                cm

b   0 1 2 3 4 5 6 7 8 9 10
                cm

c   3   4   5   6   7   8
                cm

d   11   12   13   14
                m

e   21   22   23
                m

f   30 40 50 60 70 80 90
                mm

Read each scale carefully. You need to work out what each division on the scale represents.

2 **a** Kalik's bed is 2 metres long. How many centimetres is this?
    b A fridge is 600 mm wide. How many centimetres is this?

3 Convert: **a** 5 metres into centimetres
              b 3600 centimetres into metres
              c 5000 metres into kilometres
              d 1.5 kilometres into metres.

4 **a** Convert each of these lengths into centimetres:
          i 17 m   **ii** 3400 mm   **iii** 9 m   **iv** 420 mm.
    b Which is the longest?

5 Penny measures the length of a lawn as 23.47 m.
    Write this length
    a correct to the nearest metre
    b correct to the nearest tenth of a metre
    c in centimetres.

6 Hasim is 172 cm tall.
    What is Hasim's height in metres?

7   Write
   a   2.5 m in centimetres   b   693 mm in centimetres
   c   1.2 km in metres   d   4550 m in kilometres
   e   1536 mm in metres   f   0.25 km in metres.

8   Write the following lengths in order with the shortest first.
   25 m   156 cm   2889 mm   3.8 m   0.57 m

> You need to write all the lengths in the same units before you compare them.

9   One fence post is 150 cm long.
   What length of wood, in metres, is needed to make ten of these fence posts?

10   When Liz goes to school she walks 450 m to the bus stop.
   She then has a bus journey of 1.65 km followed by a walk of 130 m.
   Work out the total distance that Liz travels to school.

> All quantities must be in the same units before you can add them.

11   This sketch shows four things on a kitchen unit.
   The milk bottle is 18 cm high.
   Estimate
   a   the height of the cup
   b   the height of the wine bottle
   c   the diameter of the saucer
   d   the height of the packet of Tasty Flakes.

> Estimate the height of the cup as a fraction of the height of the milk bottle. Then find that fraction of 18 cm.

A01   12   Kate measures the width of a wall as 135 cm.
   She sees a bookcase in a catalogue that is 1270 mm wide.
   Will the bookcase fit the wall?
   Give a reason for your answer.

13   Find 30 cm as a fraction of 4 m.

> To find one quantity as a fraction of another, first make sure they are both in the same units. Then put the first quantity over the second. Remember to simplify the fraction.

14   A rough sawn plank is 2 cm thick.
   When it is smoothed, 2.5 mm is taken off the thickness.
   What fraction of the thickness is removed?

15   There was 5 m of tape on this roll.
   Bill used 284 cm of the tape.
   What length of tape is left on the roll?

7.2 Imperial units of length

The mile is the only Imperial unit of length that is still in everyday use in the UK. Yards, feet and inches are other Imperial units of length that are used occasionally.

The relationships between these units are
   1 foot = 12 inches, 1 yard = 3 feet, 1 mile = 1760 yards.

You can convert between Imperial and metric units of length using
  5 miles ≃ 8 km,
  2 inches ≃ 5 cm.

You need to learn these.

The flow charts show how you can use these relationships.

The symbol ≃ means 'is approximately equal to'. It is used because these relationships are not exact.

Example 2

a Convert 20 miles into kilometres.
b Convert 40 cm into inches.

a 20 ÷ 5 = 4 and 4 × 8 = 32
    20 miles is about 32 kilometres.

To change miles to km, divide by 5 then multiply by 8.

b 40 ÷ 5 = 8 and 8 × 2 = 16
    40 cm is about 16 inches.

To change cm to inches, divide by 5 then multiply by 2.

Exam practice 7B

1 Write **a** 24 inches in feet **b** 12 feet in yards.

2 Find approximately
  a 10 miles in km   **b** 40 miles in km   **c** 80 km in miles
  d 12 inches in cm   **e** 40 cm in inches   **f** 110 cm in inches
  g 80 km in miles   **h** 112 km in miles   **i** 144 km in miles.

Use the flow charts above to help you.

3 Which of these distances is the greater?
  a 80 km or 60 miles **b** 1 km or 1 mile   **c** 1 inch or 1 cm

Write 80 km in miles, then you can see which is greater.

4 Arrange these lengths in order of size with the longest first.
  24 inches, 1500 mm, 100 cm, 1 foot

Write each length in the same units such as centimetres. Remember 1 foot = 12 inches.

5 The diameter of a pipe is $1\frac{1}{2}$ inches.
  1 inch is equal to 25.4 mm.
  Find the diameter of the pipe in millimetres.

$1\frac{1}{2}$ inches = $1\frac{1}{2}$ × 25.4 mm

A01

6 A road sign in France gave the distance to Paris as 84 km.
  1 km is equal to 0.621 miles.
  Jordan said 'That means the distance is more than 50 miles.'
  Is Jordan correct?
  Give a reason for your answer.

7 Use 1 mile = 1.609 kilometres to write 250 miles as a number of kilometres.

8 This chart shows the distances of air routes between some cities in Europe.
   a How far is it from Paris to Rome?
   b How far is it from London to Stockholm?
   c Work out the distance between London and Stockholm in miles.

	London				
Paris	334	Paris			
Rome	1427	1100	Rome		
Madrid	1226	1038	1338	Madrid	
Stockholm	1452	1540	2010	2580	Stockholm
Athens	2389	2086	1050	2350	2530

Distances are in kilometres

9 John said that 880 yards was further than $\frac{1}{2}$ mile. Is John right? Explain your answer.

10 An old floor board is 6 inches wide.
   The widths of new floor boards are given in millimetres.
   Decide which of these widths can be used to replace the old board.

   140 mm, 150 mm, 160 mm

   Use 1 inch = 25.4 mm.
   Give a reason for your choice.

7.3 Metric units of mass

Mass is the scientific name for the amount of matter in an object.
In everyday language we usually talk about the weight of something rather than its mass.

> In science, weight is a force caused by gravity.

The main metric units for measuring mass are the tonne (t), the kilogram (kg) and the gram (g).

The relationships between them are
   1 tonne = 1000 kilograms,
   1 kilogram = 1000 grams.

The diagram below shows how you can use these relationships to change a weight given in one unit to another unit.

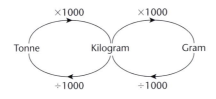

Did you know

that the exact weight of a kilogram is measured against a kilogram weight made in England out of platinum-iridium in 1889? It is kept in a sealed vault in Paris.

But over the years it's got slightly lighter. The difference is about the same as a single grain of sugar.

Example 3

a Change 500 g to a number of kilograms.
b Change 1.5 tonnes to a number of kilograms.

a 500 ÷ 1000 = 0.5
   500 g = 0.5 kg

> The arrows on the chart show you divide by 1000.

b 1.5 × 1000 = 1500
   1.5 t = 1500 kg

> The arrows on the chart show you multiply by 1000.

Exam practice 7C

1 Write down the weight shown on each set of scales.

a 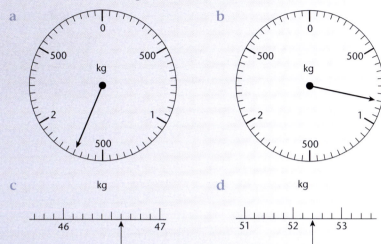 b

> You should give your answer as a single unit. In **a** use kg, in **b** use g.

c   kg   d   kg

46   47   51   52   53

2 Copy each scale and draw an arrow on your copy to show
   a   a weight of 250 g   b   a weight of 62.5 kg.

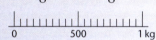
0   500   1 kg

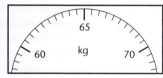

3 Write
   a   500 g in kilograms   b   1.3 t in kilograms
   c   250 kg in tonnes   d   1.35 kg in grams
   e   45 500 kg in tonnes   f   12 000 g in tonnes.

4 a   A bag of potatoes weighs 2.5 kg. How many grams is this?
   b   A crate is marked 4500 kg. How many tonnes is this?

5 Which is heavier, 1.5 kg or 2000 g?   Convert 1.5 kg into grams, then you can compare the weights.

6 Arrange the following weights in order of size with the lightest first.
   0.06 tonnes, 655 kg, 62 000 g

7 A tub contains 1 kg of margarine. A portion of margarine is 10 g.
   Work out the number of portions in the tub.

8 A stack of 500 sheets of paper weighs 2.8 kg.
   Work out the weight of 1 sheet of paper.

> Convert the weight of the stack into grams.

9 A bag of flour weighs 2.5 kg.
   David uses 500 g of this flour.
   What fraction of the flour in the bag did David use?

10 The weights of three parcels are 15.8 kg, 900 g and 48.5 kg.
    Work out the total weight of the parcels.

11 A stack of bricks weighs 2.5 tonnes.
    One brick weighs 2.5 kilograms.
    Fred used 100 of these bricks to build a wall.
    Work out the fraction of the bricks that Fred used.

7.4 Imperial units of mass

All weights in the UK are now given in metric units.
Imperial units that are used occasionally are the ton, the pound (lb)
and the ounce (oz), where 1 lb = 16 oz.

You can convert, approximately, between kilograms and pounds
using 1 kg ≃ 2.2 lb.

Example 4

a Convert 2.5 kg into pounds.
b Convert 12 lb into kilograms.

a $2.5 \times 2.2 = 5.5$
   2.5 kg $\simeq 5.5$ lb

To convert kg into pounds, multiply by 2.2.

b $12 \div 2.2 = 5.45$ correct to 2 decimal places
   12 lb $\simeq 5.45$ kg

To convert pounds into kg, divide by 2.2.

Exam practice 7D

1 Convert a 32 ounces into pounds
           b $1\frac{1}{2}$ pounds into ounces.

2 Work out, approximately,
   a 5 kg in pounds   b 500 g in pounds   c 44 lb in kilograms.

3 Which is heavier, 12 kg or 20 pounds?

Convert 12 kg into pounds (this is easier than changing pounds to kilograms).

4 Arrange the following weights in order of size
   with the heaviest first.
      2.5 kg, 11 lb, 1500 g, 64 oz

Convert these weights into the same units.
Remember that 16 oz = 1 lb.

5 Andy buys a 10 kg bag of potatoes.
   Work out its weight in pounds.

6 An old recipe asks for 2 pounds of sugar.
1 pound is approximately equal to 454 grams.
Will a $1\frac{1}{2}$ kg bag of sugar be enough?
Give a reason for your answer.

7 A bar of cooking chocolate weighs 100 g.
1 ounce is approximately equal to 28.4 grams.
Is this bar enough to give 4 ounces of chocolate?
Give a reason for your answer.

7.5 Metric units of capacity

Capacity is used to measure the volume inside a container. It is also used as a measure of the volume of liquid.

Metric units of capacity in everyday use are
the litre (*l*), the centilitre (cl) and the millilitre (ml).

The relationship between these units are
1 litre = 100 centilitres = 1000 millilitres,
1 centilitre = 10 millilitres.

> You will see 25 cl on some cans of soft drinks.

This diagram shows these relationships and how to use them.

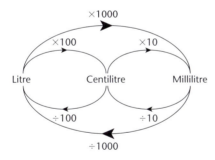

Example 5

a Convert 750 ml into litres.
b Convert 50 cl into millilitres.

a $750 \div 1000 = 0.75$
750 ml = 0.75 litres

> The arrow from ml to litres shows you need to divide by 1000.

b $50 \times 10 = 500$
50 cl = 500 ml

> The arrow from cl to ml shows you need to multiply by 10.

Exam practice 7E

1 Write down the quantity of liquid in each container.

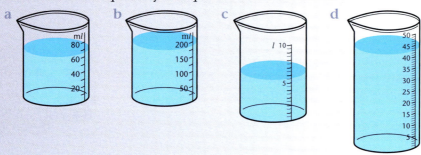

2 Write
  a 2 litres in millilitres   **b** 25 cl in millilitres
  c 500 ml in litres   **d** 900 ml in litres
  e $1\frac{1}{2}$ litres in millilitres   **f** 100 cl in litres.

3 Which holds more cola, a 30 cl can or a 250 ml bottle?

4 Arrange these capacities in order of size with the largest first.
    0.46 litres, 50 cl, 400 ml, 0.05 litres, 40 ml

5 a A container holds 10 litres of water. How many centilitres is
    this?
  b A can contains 30 cl of cola. What fraction of a litre is this?

6 A coffee jug holds 2.5 litres when full.
  A cup holds 300 ml.
  How many cups can be filled from a full jug?

7 A full bottle of weedkiller holds 0.75 litres.
  One capful holds 25 ml.
  Work out what fraction one capful is of the whole bottle.

8 A carton holds 0.80 litres of fabric conditioner.
  One wash needs 50 ml of this conditioner.
  Work out the number of washes that one carton can be
  used for.

9 A bottle holds 30 cl of medicine.
  One dose for an adult is one fiftieth of a full bottle.
  How many millilitres is an adult dose?

10 Ed must take two 5 ml spoonfuls of medicine four times
   a day.
   He has a bottle holding 200 ml. How many days will it last?

11 A jug holds $1\frac{1}{2}$ litres.
   How many glasses, each with a capacity of 125 millilitres, can be
   filled from this jug?

12 a The petrol tank of George's car holds 44 litres when full.
       How much does the dial show is in it now?

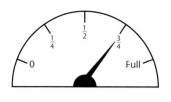

   b The petrol tank of Miriam's
      car holds 60 litres when full.
      How much does the dial
      show is in it now?

7.6 Imperial units of capacity

Gallons and pints are Imperial measures of capacity that are still used.
The relationship between them is 1 gallon = 8 pints.

You can use 1 gallon ≃ 4.5 litres to convert between gallons and
litres.

Example 6

About how many litres is 20 gallons?

$20 \times 4.5 = 90$

20 gallons ≃ 90 litres

> 1 gallon ≃ 4.5 litres,
> so 20 gallons ≃ 20 × 4.5 litres.

Exam practice 7F

1 Convert a 16 pints into gallons
           b $1\frac{1}{2}$ gallons into pints.

2 An old petrol tank holds 5 gallons.
   Approximately, how many litres will this tank hold?

3 The capacity of an oil tank is 58 gallons.
   Find the capacity of the tank in litres.

4 Does a 2 gallon petrol can hold more or less than a 10 litre
   petrol can?
   Give a reason for your answer.

5 Sandra buys 36 litres of petrol.
   How many gallons is this?

> 1 gallon ≃ 4.5 litres, so
> you need to divide 36
> by 4.5.

6 Arrange these capacities in order of size with the smallest first.
       1 litre, 10 pints, 1 gallon.

> Remember that
> 1 gallon = 8 pints.

7 Brijesh has a watering can that holds 10 gallons.
Can he get 50 litres of water into his can? Explain your answer.

8 The capacity of a watering can is marked in half gallons from 1 to 5 gallons.
A liquid fertiliser needs to be mixed with 10 litres of water.
Which mark should the watering can be filled to?
Explain whether your answer is exact or approximate.

7.7 Best buys

Many groceries come in different sized packs. Instant coffee is usually sold in jars containing different weights.
The 'best buy' or the 'best value for money' is the pack giving the lowest cost for the same weight.

> **Did you know**
> that many supermarkets show the cost per kg, so you can compare prices of different sized packs and brands?

Example 7

Which of these two jars is the better value for money?

200 g
£1.75

1 kg
£7.45

Method 1

The smaller jar holds 200 g and costs £1.75.
This is £1.75 × 5 = £8.75 per kg.

The larger jar holds 1 kg and costs £7.45.

The larger jar is better value for money.

> There are two ways of doing this.
> This method works out the *cost of the same weight* for each jar.

> 1 kg is five times 200 g.

Method 2

The smaller jar costs £1.75 for 200 g,
so £1 will buy 200 g ÷ 1.75 = 114 g to the nearest gram.

The larger jar costs £7.45 for 1 kg,
so £1 will buy 1 kg ÷ 7.45 = 1000 g ÷ 7.45
                            = 134 g to the nearest gram.

The larger jar is better value for money.

> The second method is to find the weight of coffee that the *same amount of money* will buy.

> £1 will buy more coffee in the larger jar than in the smaller jar.

A01

Exam practice 7G

1   Peppers are sold in packs of
    three or singly.
    Which is the best value?
    Give a reason for your answer.

3 for £1·11   29p each

> You will get marks for method in an
> exam even if you make a mistake.
> *You will only get those marks if you
> write down your working.*

> You can either find the cost of 3 of
> the single peppers or find the cost
> of one of the peppers in the pack.

2   Cans of cola are sold in packs of 4 cans for £2.40
    and in packs of 6 cans for £3.48.
    Which pack is better value for money?
    Give a reason for your answer.

3   These are two different bags of paper clips.
    Which bag is better value for money?
    Explain your answer.

100 clips,
£1·25   75 clips,
£1·05

> You can compare the
> number of clips per
> penny, the cost of 25
> clips or the cost of one
> clip. Use whichever
> way you find easier.

4

540 g
for £2·54   1 kg for £3·70

Which pack of tomatoes is better value for money?
Give a reason for your answer.

5   Edam cheese is sold from the delicatessen counter
    where it is priced at £3.52 per 500 g.
    Edam cheese is also sold in prepacks weighing
    200 g and costing £1.30 a pack.
    Which way of buying Edam cheese is better value?
    Explain your answer.

6   This diagram shows the costs of buying two different sized jars
    of coffee.

Brazil
COFFEE
75g
£1.20

Brazil
COFFEE
200g
£2.90

a   What is the cost of 25 g from the smaller jar?
b   What is the cost of 25 g from the larger jar?
c   Which jar is better value for money?
d   An even larger jar contains 250 g of coffee and costs £3.70.
    Is this better value for money than either of the other two?
    Give a reason for your answer.

7 Which jar is better value?

£1.50   £3

Give a reason for your answer.

8 Olive oil is sold in four different sizes:

$\frac{1}{4}$ litre 90p

$\frac{1}{2}$ litre 125p

1 litre 225p

2 litres 430p

Yoshi needs $1\frac{3}{4}$ litres of oil. What is the cheapest way of buying it?

7.8 The range in which a rounded number lies

When you round a number to a given number of decimal places, you round up when the next decimal place is 5 or more and you round down when it is less than 5.

Starting with a number that has been rounded you can work backwards to find its smallest possible value and its largest possible value.

Example 8

250 people, to the nearest 10 people, get on to a ferry.

Find **a** the smallest number that could be on the ferry,

   b the largest number that could be on the ferry.

a 245

> The smallest number that can be rounded up to 250 is 245.

b 254

> The largest number that can be rounded down to 250 is up to, but not including 255. You can only have a whole number of people, so the answer is 254.

Measurements do not take whole number values only, they can have any value in a range.

Example 9

A line is 56 mm long correct to the nearest millimetre.
a Find the range in which this length lies.
b Illustrate the range on a number line.

a The line is from 55.5 mm up to but not including 56.5 mm long.

If a mm is the length of the line then

$55.5 \leqslant a < 56.5$.

> The symbol $\leqslant$ means 'is less than or equal to' and $<$ means 'is less than'.

b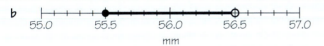

mm

> The solid circle shows that 55.5 is included in the range. The open circle shows that 56.5 is not included.

Exam practice 7H

1 The contents in this box is given to the nearest ten matches.
   a Find the least number of matches that could be in the box.
   b Find the greatest number of matches that could be in the box.

2 The weight of a bag of sand is 5.6 kg correct to 1 decimal place. Find the range of values in which the weight lies.

3 A shop made a profit of £2500 one month.
   This figure is correct to the nearest £100.
   Find the smallest possible profit.

4 The length of a room is 2.8 m correct to one decimal place.
   Write down the smallest possible length of the room.

5 A metal pin in a hinge has to have a diameter of 1.25 mm to work properly.
   This diameter is correct to 2 decimal places.
   Find the range in which the diameter must lie.

6 To the nearest 100 people, the population of a town is 130 700.
   a Find the largest possible value of the population.
   b Find the least possible value of the population.

7.9 Accuracy of answers

Measurements cannot be exact. You can assume any measurement is rounded.
For example, if you are told that a stool is 36 cm high, you can assume that this means 36 cm to the nearest centimetre.

It is reasonable to round answers to the same accuracy as the measurements in the question.

Example 10

A length of 15 m of wire is cut into 8 pieces of equal length.
Find the length of each piece.

$15 \div 8 = 1.875$

Each piece is 2 m long to the nearest metre.

> Assume 15 m is rounded to the nearest metre, so give the answer to the same degree of accuracy. It is important to say how you have rounded your answer.

This set of questions is mixed; you need to know about all the units used in this chapter.

Exam practice 7I

1 Write down the unit you would use to measure
   a   the weight of a fruit bun
   b   the weight of a spoonful of jam
   c   the length of the room you sleep in
   d   the distance between Manchester and York
   e   your height
   f   the amount of water in a tank
   g   the weight of a bus
   h   the length of your foot
   i   the length of an eyelash
   j   the amount of water in a cup
   k   the quantity of milk in a carton.

2 Convert   a   5.5 metres into centimetres
           b   1500 grams into kilograms
           c   12 cm into millimetres
           d   750 grams to kilograms
           e   4 kg to grams
           f   200 ml to litres.

3 Tony says he is 5 feet 10 inches tall.
   a   Give Tony's height in inches.
   b   Give Tony's height in centimetres.

> 1 inch ≃ 2.54 cm

4 Which is greater, a distance of 150 km or a distance of 100 miles?

> 5 miles ≃ 8 kilometres.

5 Which is heavier, 1.5 kilograms of flour or 3 lb of potatoes?

6 In January, the water in a lake was 4.6 metres deep.
In September the water in the lake was 3.8 metres deep.
What is the difference in the depth of water between January and September?
Give your answer in centimetres.

7 A full bottle contains 2 litres of lemonade.
One cup holds 75 ml.
How many cups can be filled from the full bottle?

8 A fuel tank holds 80 litres of petrol.
How many gallons is this?
Use 1 gallon ≃ 4.5 litres.

9 Alan measured the width of the space between two kitchen
units to the nearest centimetre. He wrote down 1.65 m.
a Write down the range within which the width of this space
lies.
b Alan looked at a cupboard that he wanted to fit into the
space.
The cupboard is 1651 mm wide.
Alan decided that the cupboard would not fit.
Explain why Alan could be wrong.

10 The sides of a cube are 34 mm, correct to the nearest millimetre.
The inside of a box measures 34.2 mm by 34.2 mm
correct to 3 significant figures.
Explain why the cube may not fit in the box.

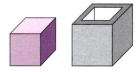

Summary of key points

- The relationships between the main metric units of length are given in this diagram.

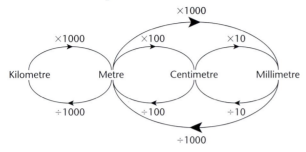

- The relationships between the main metric units of mass are given in this diagram.

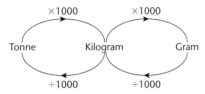

- The relationships between the main metric units of capacity are given in this diagram.

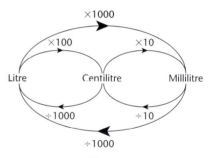

- You multiply when you convert to a smaller unit, e.g. metres to millimetres.
- You divide when you convert to a larger unit, e.g. grams to kilograms.
- You can convert between metric and Imperial units using
  5 miles ≃ 8 kilometres, 10 cm ≃ 4 inches,
  1 kg ≃ 2.2 lb, 1 gallon ≃ 4.5 litres.

Most students who get GRADE E or above can:
- convert a measurement in one unit into another unit.

Most students who get GRADE C can also:
- find the greatest and least value of a rounded number.

Glossary

Capacity	the amount that a container will hold
Imperial units	a set of units used to measure length, mass and capacity; only the mile and the pint are still officially in use in the United Kingdom
Mass	the amount of matter in an object; called weight in everyday language
Metric units	a set of units used to measure length, mass and capacity

8 Percentages

8.1 Changing percentages to fractions

The words **per cent** mean 'out of 100'.
So percentages are fractions with **denominators** of 100.

You can use this to write percentages as fractions.

> 20 per cent means 20 out of 100, or $\frac{20}{100}$.
> The symbol % is used for 'per cent'.
> So 5% means $\frac{5}{100}$, and 34% means $\frac{34}{100}$.

Example 1

Express 15% as a fraction in its lowest terms.

$$15\% = \frac{15}{100} \xrightarrow{\div 5} = \frac{3}{20} \xleftarrow{\div 5}$$

> Remember that you can simplify a fraction by dividing top and bottom by the same number.

Example 2

Express 44% as a fraction, simplifying your answer.

$$44\% = \frac{44}{100} \xrightarrow{\div 2} = \frac{22}{50} \xrightarrow{\div 2} = \frac{11}{25}$$

> You can divide by 4: $\frac{44}{100} \xrightarrow{\div 4} = \frac{11}{25} \xleftarrow{\div 4}$

8.2 Changing percentages to decimals

To express a percentage as a decimal, divide by 100 and remove the percentage sign.

Example 3

Express each percentage as a decimal. **a** 5%   **b** 44%   **c** 12.5%

a $5\% = \dfrac{5}{100} = 5 \div 100 = 0.05$

b $44\% = \dfrac{44}{100} = 44 \div 100 = 0.44$

c $12.5\% = \dfrac{12.5}{100} = 12.5 \div 100 = 0.125$

> To divide by 100, move the digits 2 places to the right (the number gets smaller).

Exam practice 8A

1 Write these percentages as fractions and simplify your answers.
   a 40%   b 80%   c 70%   d 30%   e 60%

2 Write these percentages as fractions in their simplest form.
   a 75%   b 65%   c 45%   d 35%   e 15%

3 Write these percentages as fractions in their simplest form.
   a 27%   b 42%   c 4%   d 68%   e 6%

4 Write these percentages as decimals.
   a 28%   b 81%   c 72%   d 12%   e 98%

5 Write these percentages as decimals.
   a 8%   b 1%   c 2%   d 6%   e 9%

6 Write these percentages as decimals.
   a 3%   b 67%   c 5.5%   d 12.5%   e 17.5%

7 95% of students have a job.
   What fraction of students have a job?

8 10% of the eggs in a box are cracked.
   What fraction of the eggs are cracked?

8.3 Changing decimals and fractions to percentages

To express a fraction or a decimal as a percentage you need to write it as a fraction with a denominator of 100.

You can do this easily with a decimal because the decimal places give numbers of tenths, hundredths, and so on.

For example $0.8 = \frac{8}{10} = \frac{80}{100} = 80\%,$ $0.25 = \frac{25}{100} = 25\%,$

> In each case the percentage is 100 times larger than the decimal.

> Multiply top and bottom by 10.

$0.05 = \frac{5}{100} = 5\%,$ $0.085 = \frac{85}{1000} = \frac{8.5}{100} = 8.5\%.$

> Divide top and bottom by 10.

To write a decimal as a percentage, multiply the decimal by 100 and add a percentage sign.

Example 4

Express each decimal as a percentage. **a** 0.243   **b** 1.6

a $0.243 = 0.243 \times 100\% = 24.3\%$

b $1.6 = 1.6 \times 100\% = 160\%$

> Move the digits 2 places to the left. Fill in empty places with 0s.

You can write a fraction as a percentage by writing it as a decimal first.

Example 5

Write $\frac{4}{5}$ as a percentage.

$\frac{4}{5} = 4 \div 5$
$= 0.8 = 80\%$

> Change the fraction to a decimal first by dividing the top by the bottom.

Exam practice 8B

1 Write these decimals as percentages.
  a 0.54 **b** 0.86 **c** 0.15 **d** 0.39 **e** 0.55
  f 0.04 **g** 0.01 **h** 0.06 **i** 0.09 **j** 0.02

2 Write these decimals as percentages.
  a 0.3 **b** 0.2 **c** 0.7 **d** 0.6 **e** 0.1
  f 0.35 **g** 0.035 **h** 0.925 **i** 0.175 **j** 0.405

3 Write these decimals as percentages.
  a 1.32 **b** 1.5 **c** 2.4 **d** 1.05 **e** 2.555

4 Write these fractions as percentages.
  a $\frac{1}{2}$ **b** $\frac{3}{4}$ **c** $\frac{5}{10}$ **d** $\frac{3}{5}$ **e** $\frac{1}{20}$

> First write the fraction as a decimal by dividing the top by the bottom.

5 Write these mixed numbers as percentages.
  a $1\frac{1}{2}$ **b** $1\frac{1}{10}$ **c** $1\frac{1}{5}$ **d** $2\frac{1}{4}$ **e** $1\frac{3}{4}$

6 Copy this table and fill in the blank spaces.

Fraction	Percentage	Decimal
	20%	
		0.25
	50%	
$\frac{7}{20}$		
		0.8
		0.125
	120%	

7 Work out:
   a $\frac{9}{20}$ as a percentage
   b   48% as a decimal
   c   36% as a fraction in its lowest terms
   d   56% as a fraction in its lowest terms
   e   32% as a decimal
   f $\frac{21}{25}$ as a percentage.

8 Lisa ate $\frac{2}{5}$ of a bar of chocolate.
   What percentage of the bar did she eat?

9 Jake got $\frac{17}{20}$ of the questions in a test correct.
   Find the percentage of the questions he got correct.

Answers that are not exact

Many calculations will not work out exactly when you use decimals.

Example 6

Write   **a** $\frac{1}{3}$ as a decimal correct to 3 decimal places
        b   0.3333 as a percentage correct to 1 decimal place.

   a $1 \div 3 = 0.3333 = 0.333$ correct to 3 decimal places

> Work to 4 d.p. to get your answer correct to 3 d.p.

   b $0.3333 = \frac{3333}{10000} = \frac{33.33}{100} = 33.3\%$ correct to 1 decimal place

---

Exam practice 8C

1 Write these fractions and mixed numbers as percentages correct to 1 decimal place.
   a $\frac{1}{12}$ **b** $\frac{5}{9}$ **c** $\frac{2}{15}$ **d** $\frac{3}{7}$
   e $\frac{2}{3}$ **f** $\frac{1}{6}$ **g** $1\frac{1}{3}$ **h** $1\frac{2}{9}$

2 Joy spends $\frac{5}{12}$ of her income on rent.
Correct to the nearest whole number, what percentage of her income is this?

3 A shop is having a sale:

> ### SPECIAL OFFER
> $\frac{1}{3}$ off all marked prices

Work out, to the nearest whole number, the percentage off the marked price.

4 Colin's wage is £260.
£75 is taken off his wage for tax.
a Work out the tax as a fraction of his wage.
b Change this fraction to a percentage, correct to the nearest whole number.

5 The price of a laptop computer is £450 plus $17\frac{1}{2}$% VAT.
Write $17\frac{1}{2}$% as a decimal.

> Start by writing $17\frac{1}{2}$ as 17.5.

A01

6 Stella said '33% off is the same as $\frac{1}{3}$ off.'
Is Stella correct?
Give a reason for your answer.

> **Did you know**
> that **Value Added Tax (VAT)** is a tax which is added to most things you buy?

7 a Write down the smallest of these numbers.
0.6, $\frac{2}{3}$, 65%
b Write down the largest of these numbers.
0.04, $\frac{1}{8}$, $4\frac{1}{2}$%

> Write each number as a decimal. Then you can see which is the smallest.

8 Write these numbers in order of size, starting with the smallest.
0.12, 21%, $\frac{1}{10}$, 11%

8.4 Finding a percentage of a quantity

You know how to find a fraction of a quantity.
You can find a percentage of a quantity by finding $\frac{1}{100}$ of the quantity and multiplying by the percentage needed.

Example 7

Find 3% of £12.

1% of £12 = £12 ÷ 100 = £0.12
Then 3% = £0.12 × 3
= £0.36 or 36p.

> First find 1% of £12.
> 1% = $\frac{1}{100}$,
> so 1% of £12 = £12 ÷ 100.

There are some short cuts that help you to work out common percentages of a quantity.
50% = $\frac{1}{2}$, 25% = $\frac{1}{4}$, 75% = $\frac{3}{4}$, $33\frac{1}{3}$% = $\frac{1}{3}$, 10% = $\frac{1}{10}$

> It is useful to remember these percentage facts.

Example 8

Find **a** 10% of 80 g   **b** 25% of £48   **c** $33\frac{1}{3}$% of 18 cm.

a   10% of 80 g = $\frac{1}{10}$ of 80 g = 80 g ÷ 10 = 8 g

b   25% of £48 = $\frac{1}{4}$ of £48 = £48 ÷ 4 = £12

c $33\frac{1}{3}$% of 18 cm = $\frac{1}{3}$ of 18 cm = 18 cm ÷ 3 = 6 cm

Exam practice 8D

1 Find:
  a 25% of £80   **b** 10% of 30p   **c** 20% of £6
  d 50% of £86   **e** 60% of £40   **f** 15% of £20.

2 Find:
  a 5% of £360   **b** 200% of £250   **c** 150% of £300
  d 30% of 50 kg   **e** $33\frac{1}{3}$ of 600 grams **f** 75% of 40 metres.

3 Find:
  a 40% of £2.50   **b** 10% of £1.20   **c** 70% of 1.2 km.

4 Roger buys a book on the Internet priced £7.80.
The delivery charge is 5% of the price. Find the delivery charge.

5 Holly sells her house for £250 000.
The Estate Agent charges 2% of the sale price.
How much does the estate agent charge?

6 An electricity bill is £84.50. VAT of 5% is added to the bill.
How much is the VAT?

7 A mathematics text book has 200 pages.
10% of the pages have diagrams on them.
Work out the number of pages that do *not* have diagrams on them.

8 A gas bill is £46.20. VAT of 10% is added to the bill.
Work out the VAT.

9 There are 500 students at a college. 40% of them walk to college.
How many students walk to college?

10 Mike is paid 2% commission on everything he sells.
How much commission does he earn on sales of £14 500?

Class discussion
You can find 10% of a quantity by dividing by 10.
How can you use this fact to find each of these?
      5%
    15%
    $2\frac{1}{2}$%
    $7\frac{1}{2}$%
  $17\frac{1}{2}$%

11 A chicken farmer buys 1700 chickens.
10% of these chickens die.
How many chickens are left?

12 I owe £178.20 on my credit card.
I must pay whichever is larger: £5 or 5% of the amount I owe.
Work out the amount I must pay.

8.5 Using a calculator

You can use a calculator to help with percentage calculations.

Example 9

Find, to the nearest kilogram, 45% of 251 kg.

1% of 251 kg = 251 kg ÷ 100 = 2.51 kg.

First find 1% of the quantity. then multiply by 45.

So 45% of 251 kg = 45 × 2.51 kg
= 112.95 kg
= 113 kg correct to the nearest kg.

You can use a calculator to find the answer in one step by pressing
2 5 1 $×$ 4 5 $\%$ $=$.

Exam practice 8E

Give answers correct to the nearest penny, or correct to 1 decimal place.

1 Find:
   a 83% of £280   b 2.5% of £830   c 17.5% of £14.25
   d $117\frac{1}{2}$% of £245   e 2.3% of £2465   f 4.27% of £589
   g 36.7% of 454 g   h 16.7% of 21 cm   i 23% of 650 people
   j 110% of £40   k 103% of 65 miles   l 130% of 5 minutes.

Find 1%, then multiply by the percentage you want.

Write $117\frac{1}{2}$ as 117.5.

2 The floor area of a supermarket is 1500 square metres.
8.5% of this floor area is used for selling clothes.
Find the floor area used for selling clothes.

3 Jack buys a printer priced at £89.90 plus VAT.
VAT is $17\frac{1}{2}$% of the price. Work out the VAT.

4 Alex earns £18 000 a year.
24.5% of this is deducted for tax, national insurance and pension.
Work out the amount of money deducted.

5   A builder gives an estimate for a job of £2500 plus VAT at $17\frac{1}{2}$%.
    Work out the VAT.

6   Peggy pays 6% of her income towards her pension.
    Her income last year was £18 550.
    Work out her pension contribution.

7   These are some of the results from the 2001 national census.
    The population of the UK was 58 789 194 people.

    a   21% of the population were over 60 years old.
        How many people were over 60 years old?

    b   8.6% of the population lived in Scotland.
        How many people lived in Scotland?

8   A garage bill cost £104 plus VAT at $17\frac{1}{2}$%.

    a   Work out the VAT charged.

    b   What is the total to be paid?

9   A store takes $2\frac{1}{2}$% off the price when you pay with cash.
    The marked price of a settee is £830.
    Gerry paid cash.

    a   Work out how much was taken off the price.

    b   How much did Gerry pay?

10 There are 580 students in a school. 57% of the students are
    girls.
    How many of the students are boys?

 11 Dennis said "If two of us buy rail tickets we will save 60%."
    Explain why he is wrong.

30% OFF
ALL FARES

8.6 Interest

Interest is what you are paid when you put money into a savings
account.
This sum is a percentage of the amount in the account and it is
usually paid yearly.
The percentage paid is called the **interest rate**.

If the interest rate does not change and the interest is withdrawn or
paid into another account it is called **simple interest**.

When you borrow money, you have to pay interest on the amount
borrowed.

Example 10

Nia has £3500 in a savings account that pays a 4.5% p.a. interest rate.
How much simple interest does she receive in 3 years?

Interest for 1 year = 4.5% of £3500

    1% of £3500 = £3500 ÷ 100 = £35

    so 4.5% of £3500 = 4.5 × £35

            = £157.50

Interest for 3 years = 3 × £157.50 = £472.50

> p.a. is short for '**per annum**' which means 'each year'.

Exam practice 8F

1 Work out the simple interest for one year on:
   a £500 at 3% p.a.   b £250 at 5% p.a.
   c £400 at 3.5% p.a.   d £2000 at 4.25% p.a.

2 Work out the simple interest on:
   a £400 at 4% p.a. for 2 years   b £5000 at 3% p.a. for 3 years
   c £2500 at 5% p.a. for 4 years   d £4500 at $2\frac{1}{2}$% p.a. for 3 years.

3 A building society pays 2.5% a year interest on a savings account.
   Fiona invests £4500 for 1 year.
   How much interest does she earn?

4 David put £30 000 into a savings account.
   The interest for the first year was 5.5%.
   Find the amount in the account at the end of the first year.

5 A bank pays 0.1% p.a. interest on money deposited in a current account.
   There is £540 in my account.
   It is untouched for 1 year.
   How much interest do I get?

6 Rob puts £5000 in a savings account.
   He gets interest of 4% p.a.
   He takes the interest out at the end of each year.
   Work out how much he takes out over 5 years.

7 Shelly takes out a loan to buy a secondhand car costing £5000.
   She is charged 15% interest.
   Work out the amount of interest she pays.

8 Lisa takes out a loan of £10 000.
   She is charged 20% interest.
   Work out how much she has to pay back.

9   Zia takes out a loan of £1000.
    He pays £90 a month for 12 months to repay the loan.
    a   How much more than £1000 does Zia pay back?
    b   Work out the extra that Zia pays back as a percentage of his
        loan.

10 Pierre buys a music centre priced £550 on credit.
    The interest charged is 12% of the price.
    a   Work out how much Pierre has to pay back.
    b   He pays it back over 8 months in equal amounts.
        How much does he pay each month?

8.7 Finding one quantity as a percentage of another

You know how to find one quantity as a fraction of another.
You can find one quantity as a percentage of another by finding it
first as a fraction, then changing the fraction to a percentage.

Example 11

Find: **a** £13 as a percentage of £80   **b** 45p as a percentage of £8.36.

a   £13 as a fraction of £80 = $\frac{13}{80}$ = 0.1625

0.1625 × 100% = 16.25%

So £13 is 16.25% of £80.

> First express £13 as a fraction of £80, that is divide 13 by 80. Next convert this fraction into a decimal. Multiplying a decimal by 100 turns it into a percentage.

b   45p as a fraction of £8.36 = $\frac{45}{836}$ = 0.0538…

0.0538 × 100% = 5.4% correct to 1 d.p.

So 45p = 5.4% of £8.36 correct to 1 d.p.

> Change £8.36 into pence so that both quantities are in the same units.

Exam practice 8G

1   Write:
    a   £2.40 as a percentage of £80
    b   45 cm as a percentage of 100 cm
    c   2 ml as a percentage of 20 ml
    d   15p as a percentage of £1.50.

> Give any answer that is not exact correct to the nearest whole number or to one decimal place.

> Change £1.50 to 150p.

2 Find:
   a 36 marks as a percentage of 60 marks
   b 27p as a percentage of £1.08
   c £4.50 as a percentage of £4
   d 6 seconds as a percentage of 60 seconds
   e 120 g as a percentage of 1 kg
   f 33 cm as a percentage of 1 metre
   g 400 metres as a percentage of 1 kilometre.

> Write both quantities in the same units.

3 Twenty cars went through an MOT one day. Six of these cars failed.
   Work out the percentage that failed.

4 A pair of shoes is in a sale. The shoes were originally priced £38.50.
   Find the discount as a percentage of the original price.

> **SALE**
> £10 off all marked prices

5 Three hundred children applied for a place at a school.
   Only 90 children got places.
   What percentage of these 300 children got places?

6 Peaches cost 24p each yesterday.
   Today, peaches cost 27p each.
   Work out the cost of a peach today as a percentage of yesterday's cost.

7 Jason ordered groceries on the Internet.
   The cost of the groceries was £64 and there was a £5 delivery charge.
   Find the delivery charge as a percentage of the cost of his groceries.

8 In a survey of 2000 people, 480 said they did not own a car.
   What percentage of the people surveyed said they did not own a car?

9 Kate put £460 into a savings account.
   After two years, there was £502 in the account.
   Find £502 as a percentage of £460.

10 Valda paid £750 for an income bond.
    She got £26.25 income from this bond.
    Find the income as a percentage of the money Valda paid.

11 Todd bought a motorcycle for £1250.
    He sold it eighteen months later for £825.
    a Work out how much he lost.
    b Express this loss as a percentage of what he originally paid.

12 There are 120 boys and 130 girls in Year 11 at Stanton School.
    Work out the percentage of the year group that are
    a girls   b boys.

13 A club has 64 members. The are 24 male members.
Work out the percentage of the members that are female.

14 Students in Year 11 were asked if they had a job.
This table shows the answers:

Students in Year 11

	Job	No job
Boys	115	10
Girls	135	15

a Work out the percentage of boys without a job.
b Find the percentage of girls without a job.
c Jake said that fewer than 10% of Year 11 students do not
have a job.
Is Jake correct? Give a reason for your answer.

A01

8.8 Percentage increase and decrease

**A percentage used to describe an
increase or a decrease is always a
percentage of the quantity *before*
it is changed.**

Percentages are often used to describe
increases and decreases.
For example, 'Average council tax increases
are 8% this year.' 'Computer prices have
decreased by 20% over the last 3 years.'

Example 12

A bus fare is 80p now, and will increase by 5% next month.
Find the new fare.

This means that the
increase is 5% of 80p.

1% of 80p = 80p ÷ 100 = 0.8p
so 5% of 80p = 5 × 0.8p = 4p

The new fare is 80p + 4p = 84p.

When a quantity is decreased by a percentage, you can find the
decreased quantity in the same way.

Example 13

A new car cost £8500. In the first year it lost 20% of its value. Find its
value after one year.

Its value decreased by
20% of £8500.

1% of £8500 = £8500 ÷ 100 = £85
so 20% of £8500 = 20 × £85 = £1700
Value of the one-year-old car = £8500 − £1700
= £6800

Exam practice 8H

1 Tim's weight increased by 12% between his fifteenth and sixteenth birthdays.
  He weighed 65 kg on his fifteenth birthday.
  How many kilograms did he gain?

2 Grace's water rates are 6% more this year than they were last year.
  She paid £510 last year.
  How much more does Grace have to pay this year?

3 A man's suit has a marked price of £125.
  The marked price is reduced by 20% in a sale.
  How much is taken off the price?

4 Last year, 550 children went to the local school.
  This year there are 8% fewer children at the school.
  How many fewer children are there this year?

5 There are 800 pupils in the school this year.
  Next year there will be 40 more pupils in the school.
  Work out the percentage increase in the number of pupils.

6 Misty used to earn £280 a week.
  She now works fewer hours and earns £245 a week.
  Find: a the reduction in her earnings
         b the percentage decrease in her earnings.

7 A factory employs 220 workers.
  Next year the work force will be increased by 15%.
  How many more workers will be employed next year?

8 At the normal rate a machine fills 600 cans an hour.
  At its fastest rate it fills 15% more cans an hour.
  How many more cans per hour does it fill at the fastest rate?

9 A secondhand car dealer buys a car for £4225.
  He increases the price by 20% to sell it.
  How much profit does he make?

10 Anne sees a dress in a shop priced £56.
   In the sale all prices are reduced by 15%.
   How much does Anne save?

11 A bathroom suite is marked at £850.
   VAT (value added tax) is added at $17\frac{1}{2}$% of the marked price.
   Work out the amount Stan has to pay to buy this suite.

12 All prices are reduced by 10% in a sale.
   a What is the sale price of a coat marked £40?
   b What is the reduction on a suit marked £85?

Give any answer that is not exact correct to the nearest whole number or to one decimal place.

Make sure that you know whether you are asked to find just the increase or decrease or the new amount.

Decide if you are asked to find a percentage of a quantity or one quantity as a percentage of another.

Profit means the difference between the selling price and the buying price.

13 There were 36 reported cases of measles last year in the local
    health area.
    This year the number of reported cases has dropped by 15%.
    How many cases have been reported this year?

14 Kylie had a crop of 25 kg of tomatoes last year.
    This year she had a crop of 30 kg of tomatoes from the same
    number of plants.
    Find the percentage increase in her crop.

15 Khalid bought 120 bottles of water for £54.
    He sold them for 60p each.
    Find his percentage profit.

> The percentage profit is the profit as a percentage of the amount originally paid.

16 Sandra earns £4 an hour for her Saturday job.
    She gets a raise of 5%.
    Work out her new hourly rate.

17 Tim paid £95 000 for a house.
    Two years later he sold it for £106 400.
    Work out the percentage increase in its value.

18 A bookshop has this offer.

---

3 FOR 2
BUY 3 BOOKS AND GET THE CHEAPEST ONE FREE

Raj bought 3 books.
The prices were £5.99, £7.55, £6.99.
Find his percentage saving.

19 Rob has carved a model of a horse.
    The piece of wood he started with weighed 9 kg.
    He cut away 72% of the wood.
    How much did the completed horse weigh?

20 Russ bought a picture for £4000. After one year it had gone
    up in value by 20%. The following year it went down in value
    by 20% of its value at the beginning of that year. What was it
    worth at the end of the second year?

> Read the question carefully. Do not jump to conclusions.

21 Peter Brown gets a basic weekly wage of £265 plus commission
    at 2% of the value of the goods he sells. Last week he sold goods
    to the value of £8500.
    a   How much commission did he earn?
    b   What was his total pay for the week?

Summary of key points

- You can write a percentage as a fraction with a denominator of 100.
- You can write a percentage as a decimal by dividing it by 100.
- You can write a decimal as a percentage by multiplying it by 100.
- You can write a fraction as a percentage by first writing the fraction as a decimal (divide the top by the bottom) then multiplying the decimal by 100.
- You find a percentage of a quantity by first finding 1% (divide the quantity by 100) then multiplying by the percentage you want.
- You find one quantity as a percentage of another by putting the first quantity over the second and multiplying by 100 (make sure that both quantities are in the same units).
- A percentage used to describe an increase or decrease is always a percentage of the quantity before it is changed.
- You need to read percentage questions carefully to make sure that you know whether you are being asked to find a percentage of a quantity or one quantity as a percentage of another.

Most students who get GRADE E or above can:
- change between simple fractions, decimals and percentages
- find the percentage of a quantity
- work out simple interest questions.

Most students who get GRADE C can also:
- calculate percentage increase and decrease.

Glossary

Denominator	the bottom number in a fraction
Interest	money that is paid on savings by banks (or by you if you borrow money)
Interest rate	the percentage of the money you borrow or lend that is paid as interest
Per annum	each year
Per cent	each hundred
Simple interest	interest that stays the same each year
VAT	value added tax

9 Ratio and proportion

9.1 Ratio

Ratio is a way of comparing quantities.

The label on a bottle of squash says
'Mix 1 part of squash with 4 parts of water.'

This compares the quantities of squash and water needed to make a drink.

It is called the **ratio** of the amount of squash to the amount of water.
You can write this as 'the ratio of squash to water is 1 to 4' or simply
as squash : water = 1 : 4.

> The symbol ':' means 'to'

Simplifying ratios

Ratios are easier to work with when the numbers are as small as possible.
You can reduce the size of the numbers in a ratio by simplifying.
Simplifying ratios is very similar to simplifying fractions.

Example 1

Write this instruction as a ratio in its simplest terms:

'Mix 200 ml of syrup with 500 ml of water'

> You can simplify the ratio 200 ml : 500 ml by leaving out the units.
> You can always leave out the units as long as they are the *same for both quantities*.

$$\text{Syrup : water} = 200\,\text{ml} : 500\,\text{ml}$$
$$= 200 : 500$$
$$\div 100 \overset{\frown}{} \div 100$$
$$= \quad 2 : 5$$

> You can now simplify 200 : 500 by dividing each number by 100.

When you simplify ratios you must make sure that both quantities are in the same units.

Example 2

Simplify the ratio 2 m : 30 cm.

$$2 \text{ m} : 30 \text{ cm} = 200 \text{ cm} : 30 \text{ cm}$$
$$= 200 : 30$$
$$= 20 : 3$$

> You must write both lengths in centimetres (or metres). Then you can leave out the units.

> You can reduce the sizes of the numbers by dividing each by 10.

Exam practice 9A

1. Simplify these ratios.
   a 4 : 6   b 10 : 30   c 50 : 200
   d 8 : 20   e 400 : 80   f 250 : 90

> Look for a number that divides exactly into both parts of the ratio.

2. Simplify these ratios.
   a 2 cm : 8 cm   b 96p : 40p   c 20 ml : 5 ml
   d 32p : 96p   e £45 : £18   f £350 : £70

3. Simplify the ratios.
   a $3 \text{ m} : 1\frac{1}{2} \text{ m}$ b 0.9 mm : 1.8 mm   c $4\frac{1}{2}$ inches : 9 inches
   d $3 \text{ cm} : 1.5 \text{ cm}$ e $3\frac{1}{3} \text{ m} : 5 \text{ m}$ f 1.5 cm : 2.5 cm

> For **a** you need to double both parts. For **b** you can start by multiplying both parts by ten.

4. Simplify these ratios.
   a £4 : 75p   b 48p : £2.88   c £1.50 : 75p
   d 1 m : 20 cm   e 2 kg : 500 g   f 20 kg : 500 g

> Make sure both quantities are in the same units before you simplify.

5. Which of these ratios is equivalent to 2 : 3?
   a 6 : 8   b 10 : 15   c 9 : 6   d 12 : 18

A01 6. Clive said that the ratio 25 : 15 is equivalent to 50 : 30.
   Is Clive correct? Give a reason for your answer.

A01 7. Edna said that the ratio $\frac{1}{2} : \frac{1}{4}$ is equivalent to 1 : 2.
   Is Edna correct? Give a reason for your answer.

8. Find the missing number in this ratio, □ : 4 = 9 : 12.

9. Write these ratios in their simplest form:
   a 1 mm : 4 cm   b 55p : £2   c 45 minutes : 2 hours

10.
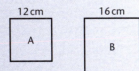

Find the ratio of the length of the side of square A to the length of the side of square B.

> Make sure that the numbers in the ratio are in the same order as the words.

11. Find the ratio of the length of this pool to its width.

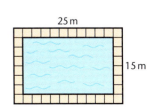

12. Last year John earned £40 000 and spent £25 000.
    Find the ratio of the money he spent to his earnings.

13 Arshi walks 2 km to school and David cycles 4 km.
Find the ratio of Arshi's distance to David's distance.

14 A road that is 12 km long is shown by a line on a map that is
4 cm long.
Find the ratio of the length of the line on the map to the length
of the road.

15 There are 12 boys and 8 girls in a playground.
  a Find the ratio of the number of boys to the number of girls.
  b Find the ratio of the number of girls to the number of
     children.

Read the question
carefully. Make sure
you understand which
quantities you need.

16 A 50 ml bottle of liquid plant food is diluted with 10 litres of
water.
Find the ratio of liquid food to water.

9.2 Using ratios

A **map ratio** tells you the ratio of distances on the map to distances
on the ground.

Example 3

A path on this map is 3 cm long.
How long is the path on the ground?

The scale on this map
is 1 : 10 000. This is the
map ratio.
It means that a line on
the map shows a line
on the ground that is
10 000 times longer.

The path on the ground is 3 × 10 000 cm long.
3 × 10 000 cm = 30 000 cm
              = 300 m
So the path is 300 m long.

Scale drawings are like maps. The drawing is a small version of the
real thing.
The scale tells you the ratio of lengths on the drawing to the real lengths.

Example 4

This is a scale drawing of a garden. It is drawn on a 1 cm grid.
The scale is 1 : 300.
Find the width of the garden.

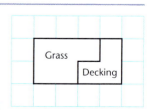

The garden is 2 cm wide on the drawing,
so the width of the real garden is 300 × 2 cm = 600 cm
                                              = 6 m.

The scale tells you
that any length on the
ground is 300 times as
long as the length on
the drawing.

A **gear ratio** compares the number of teeth on the wheels.

Example 5

a Find the gear ratio of these wheels.
b The big wheel turns once. How many times does the small wheel turn?

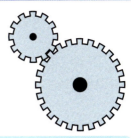

> **a** The smaller wheel has 12 teeth.
> The larger wheel has 24 teeth.
> The gear ratio is 12 : 24 = 1 : 2
>
> **b** When the big wheel turns once, the small wheel turns twice.

When the big wheel turns once, 24 teeth are engaged. The small wheel has to turn twice to engage 24 teeth.

Exam practice 9B

1 This is part of a street map.
The map ratio is 1 : 5000.
Abbey road is 2 cm long on the map.
Rick walks from one end of Abbey Road to the other.
How far does he walk? Give your answer in metres.

1 metre = 100 cm

2 A street on a map is 4 cm long.
The real street is 1 kilometre long.
Find the map ratio.

This means that you have to find the ratio 'length on map : length on ground'. Remember that 1 km = 100 000 cm.

3 The map ratio of a map is 1 : 10 000.
The distance between two places on the map is 20 cm.
How far apart are they in real life?

4 The scale of a map is 1 : 10 000.
  a A path on the map is 1 cm long.
    How long is the path on the ground? Give your answer in metres.
  b The distance between Coryton and Whatley on the map is 12 cm.
    Work out the actual distance between Coryton and Whatley.

5 The scale of a map is 1 : 50 000.
The distance between Woodburn and Axe is 9 cm on the map.
What is the actual distance between these two places?
Give your answer in kilometres.

6 Two towns are 2 km apart.
On a map they are 8 cm apart.
Work out the map ratio.

7 This is a map of an island.
It shows 4 villages.
  a Measure the distance between A and B.
  b Work out the distance between A and B on the ground. Give your answer in kilometres.

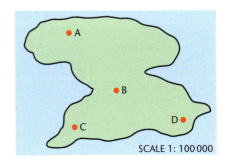
SCALE 1 : 100 000

8 The distance between London and Manchester is 320 km.
The distance between them on a road map is 16 cm.
Work out the scale as a ratio. Simplify your answer.

9 The scale of an Ordnance Survey map is 1 : 50 000.
   a What distance does 1 cm on the map represent?
   b What distance, in kilometres, does 10 cm on the map represent?
   c Two villages are 10 km apart.
     How far apart are they on the map?

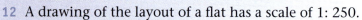

10 Julia draws the floor plan of her lounge on a 1 cm grid.
   a How long is the line from A to B on her drawing?
   b The distance from A to B in the lounge is 450 cm.
     Work out the scale used. Simplify your answer.

11 A model of a building is 12 cm high.
The ratio of the height of the actual building
to the height of the model is 200 : 1.
How high is the building?

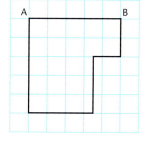

12 A drawing of the layout of a flat has a scale of 1 : 250.
   a How many metres does 4 cm on the drawing represent?
   b A wall on the drawing is 16 cm long.
     Work out how long the actual wall is.

13 Stan makes a model of a railway engine.
The scale of the model is 1 : 72.
   a A wheel on the model is 1.5 cm wide.
     Find the width of the real wheel on the engine.
   b The engine is 7.2 m long. How long is the model?

14 These are two gear wheels.
Wheel A has 16 teeth.
Wheel B has 24 teeth.
Work out the ratio of the number of teeth on B to the number of teeth on A. Give your answer as simply as possible.

15 The little wheel has 8 teeth.
The big wheel has 32 teeth.
   a Find the gear ratio.
   b The big wheel turns once.
     How many turns does the little wheel make?
   c The little wheel turns 8 times.
     How many turns does the big wheel make?

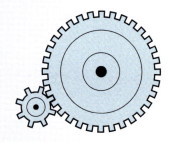

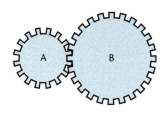

The gear ratio is 'number of teeth on the little wheel: number of teeth on the big wheel.'

16 The gear ratio of these two gears is $2:5$.

    a The big wheel turns twice.
       How many turns does the little wheel make?

    b The little wheel turns 20 times.
       How many turns does the big wheel make?

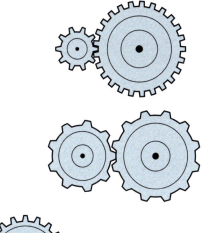

17 The gear ratio of these gears is $3:4$.

    a The small wheel turns 6 times.
       How many turns does the big wheel make?

    b The big wheel turns 3 times.
       How many turns does the small wheel make?

18 **a** Find the gear ratio of A to B.

    b Find the gear ratio of B to C.

    c What happens to wheel C when A turns once?

19 The ratio of lengths on a model boat to those on the actual boat
is $3:50$.
The real boat is 500 centimetres long.
Work out the length of the model.

20 A photograph is enlarged.
The ratio of the length of the bigger photo to the length of the
original is $5:2$.
The original photo is 10 cm long.
What is the length of the bigger photo?

9.3 Proportion

To make a potting compost you need to mix
2 parts sand with 3 parts earth.

This means that the ratio sand : earth $= 2:3$.

You can choose to mix 2 buckets of sand with 3 buckets of earth,
or   4 buckets of sand with 6 buckets of earth,
or   1 bucket of sand with $1\frac{1}{2}$ buckets of earth.

You can vary the amount of sand and earth, but they must always be
in the ratio $2:3$.

When two quantities are in the same ratio, they behave in the same
way.
If one quantity is doubled, so is the other,
If one quantity is halved, so is the other, and so on.

Quantities that are related in this way are **proportional**.

Many familiar situations involve quantities that you can assume are proportional.

> The mathematical name is **directly proportional**.

Example 6

40 envelopes cost £5.60.
Find the cost of   **a** one envelope   **b** 50 envelopes.

> You can assume that the number of envelopes is proportional to their cost.

   a   The cost of one envelope is $\frac{1}{40}$ of £5.60
   $= £5.60 \div 40 = £0.14$
   $= 14\text{p}$

   b   The cost of 50 envelopes is $50 \times 14\text{p} = 700\text{p}$
   $= £7$

Exam practice 9C

1   The cost of 1 kg of sugar is 90p.
    What is the cost of 12 kg of sugar?

2   The cost of 1 kg of mushrooms is £3.30.
    Find the cost of 2.4 kg of mushrooms.

3   Six pens cost £7.20.
    Work out the cost of one pen.

4   A man walks steadily for three hours and covers 13 km.
    How far does he walk in one hour?

5   8.6 square metres of carpet cost £71.38.
    What is the cost of 1 square metre?

6   In one hour an electric fire uses 1.5 units.
    Find how many units it uses in $\frac{1}{2}$ hour.

7   It costs 4.8p to run a fridge for 3.2 hours.
    a   Work out the cost of running the fridge for one hour.
    b   How long will the fridge run for on 1p?

8   2 litres of paint will cover 24 square metres.
    a   What area will 1 litre cover?
    b   What area will a 5 litre can cover?

9   Lisa earns £22 in 4 hours.
    a   How much will she earn in 1 hour?
    b   How much will she earn in 36 hours?

9.4 Using proportion

You can use proportion to solve problems by finding one unit of a quantity.

Example 7

Jane pays £35.70 for 3.5 m of carpet.
a How much does 5 m of the same carpet cost?
b How many metres of this carpet will £50 buy?

a 1 m of carpet costs £35.70 ÷ 3.5 = £10.20
   5 m of carpet costs £10.20 × 5 = £51

b £1 buys 3.5 ÷ 35.70 m = 0.09803…m
   £50 buys 0.09803… × 50 m = 4.901… m
   　　　　　　　　　= 4.9 m correct to 1 d.p.

> The two quantities are money and length. You have two lengths, so first find the cost of one unit of length – this is 1 m of carpet.

> You know two sums of money, so find the length that £1 buys.

Exam practice 9D

1 A hiker walks at a constant speed.
He walked 16 km in 4 hours.
Find how long it took him to walk 12 km.

> First find how long it takes to walk 1 km.

2 An electric fire uses 7.5 units in three hours.
  a How many units does it use in five hours?
  b How long does the same electric fire take to use 9 units?

> First work out how many units it uses in one hour.

> For this part you need to start by finding how long it takes to use one unit.

3 A taxi journey of 400 miles costs £96.
Find the cost of travelling 255 miles.

> You can assume that distance travelled and cost are proportional.

4 It costs £346.50 to surface a drive of area 60 m².
Work out how much it costs to surface a drive of area 46 m².

5 A machine fills 720 cartons in 6 hours.
How many cartons can it fill in five hours?

6 The tension in a knitting machine is set to 55 rows to give 10 cm.
How many rows should be knitted to give 12 cm?

7 The cost of hiring scaffolding is proportional to the time it is hired for.
It costs £378 to hire scaffolding for 42 days.
Work out how much it costs to hire the same scaffolding for 36 days.

8 Percy pays £8.10 for a 3 m length of skirting board.
How much will 7 m of the same skirting board cost?

9   Ernie is quoted £68 for two tyres.
    He decides he needs 5.
    How much will they cost him?

10 This is a recipe for a pizza to serve 4 people.
    a   What weight of tinned tomatoes is needed
        for a pizza to serve 12 people?
    b   How much milk is needed for a pizza for
        6 people?

Pizza

145 g of plain flour
90 millilitres of milk
115 g of tinned tomatoes

11 A recipe for 16 flapjacks needs 4 oz of margarine and 6 oz of
    rolled oats.
    a   How much margarine is needed to make 48 flapjacks?
    b   What weight of rolled oats is needed to make 12 flapjacks?

12 A recipe for 24 Welsh cakes needs 200 g of flour, 80 g of
    margarine and 100 g of sugar.
    a   How much sugar is required to make 36 Welsh cakes?
    b   How much margarine is needed to make 18 Welsh cakes?

13 This is a recipe for macaroni cheese sauce to serve eight.
    Write out the quantities of each ingredient to make this sauce
    for six people.

Macaroni Cheese

40 g flour
40 g margarine
200 g cheese
600 millilitres milk

14 The instructions for making a model lists the following
    materials:

Quantity	Costs
300 cm tape	1 m costs 76p
75 nails	250 nails cost £1.25
20 ml glue	100 ml costs £2.50
150 cm wood beading	2 m costs £1.08
2 sheets A4 card	a pack of 5 sheets costs £3.25

Find the cost of making this model as accurately as possible,
then give your answer correct to the nearest penny.

9.5 Division in a given ratio

Ratios and fractions are closely related.

For example, the ratio
    1 part of squash : 4 parts of water
tells you that squash is $\frac{1}{5}$ of the mixture.

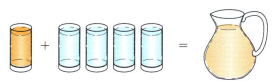

So in 10 litres of mixture, squash is $\frac{1}{5}$ of 10 litres and water
is $\frac{4}{5}$ of 10 litres.

$\frac{1}{5}$ of 10 = 10 ÷ 5 = 2   and $\frac{4}{5}$ of 10 = 4 × 2 = 8

So there are 2 litres of squash and 8 litres of water.

Finding the sizes of parts in this way is called **division in a given
ratio**.

Example 8

Divide £50 in the ratio 2 : 3.

One part is $\frac{2}{5}$ of £50 and the other part is $\frac{3}{5}$ of £50.

$\frac{1}{5}$ of £50 = £50 ÷ 5 = £10.

So one part is 2 × £10 = £20 and the other part is 3 × £10 = £30.

Check: £20 + £30 = £50.

This means you have to find sums of money that add up £50 and that are in the ratio 2 : 3.

You do this by finding each part as a fraction of the total.
Add the two numbers in the ratio to get the denominator.

More than two quantities can be compared to give a ratio.

Example 9

A company gives a bonus of £30 000 to a team of three employees.
The bonus is divided in the ratio of their earnings.
Dave earns £25 000, James earns £20 000 and Nicola earns £30 000.
Work out the bonus that each of them gets.

The ratio of Dave's to James' to Nicola's earnings

$\quad$ = £25 000 : £20 000 : £30 000

$\quad$ = 25 000 : 20 000 : 30 000

$\quad$ = 25 : 20 : 30

$\quad$ = 5 : 4 : 6

Dave gets $\frac{5}{15}$, James gets $\frac{4}{15}$ and Nicola gets $\frac{6}{15}$.

$\frac{1}{15}$ of £30 000 = £2000.

So Dave gets 5 × £2000 = £10 000, James gets 4 × £2000 = £8000

and Nicola gets 6 × £2000 = £12 000.

First work out the ratio of their earnings and simplify it as much as possible.

Add the numbers in the ratio to get the denominator of the fraction.

Check: 10 000 + 8 000 + 12 000 = 30 000.

Exam practice 9E

1 a Divide 80p in the ratio 3 : 2.
$\quad$ b Divide 32 cm in the ratio 3 : 5.
$\quad$ c Divide £45 in the ratio 4 : 5.
$\quad$ d Divide £20 in the ratio 1 : 7.

2 Soujit and Tom share 40 peanuts between them in the ratio 3 : 5.
$\quad$ How many do they each get?

3 Mary is 10 years old and Eleanor is 15 years old.
$\quad$ Divide 75p between them in the ratio of their ages.

First find the ratio of their ages.

4 There are 30 pupils in a class.
$\quad$ The ratio of the number of boys to the number of girls is 7 : 8.
$\quad$ How many girls are there?

5 The total amount of money raised at a school fair was £1029.
$\quad$ This money was divided between the school and a local charity
$\quad$ in the ratio 5 : 2.
$\quad$ The school got the larger amount.
$\quad$ How much did the local charity receive?

6 Divide £24 between Joe and Ellen so that Ellen gets twice as much as Joe.

Ellen gets twice as much as Joe means that £24 is divided in the ratio 2 : 1.

7 School shirts are made from cotton and polyester in the ratio 3 : 2 by weight.
A shirt weighs 120 g.
How much polyester is there in it?

8 A plank of wood is 3 metres long.
It is divided in the ratio 2 : 3.
Work out the length of the longer piece.

9 A fruit punch is made by mixing mango juice and pineapple juice in the ratio 2 : 7.
   a How much mango juice is there in 18 litres of this punch?
   b Work out the quantity of pineapple juice needed to make 27 litres of punch.

10 a Divide 80 kg in the ratio 2 : 3 : 5.
   b Divide £40 in the ratio 1 : 2 : 2.

11 A shortbread mixture is made by mixing flour, sugar and butter in the ratio 4 : 2 : 3.
How much butter is there in 360 grams of mixture?

12 Pete, Zoe and Sara divide £550 between them in the ratio of their ages.
Pete is 15 years old, Zoe is 15 years old and Sara is 20 years old.
How much do each of them get?

13 Divide 260 g in the ratio 5 : 8.

14 Clara and Di are given £5.70.
They divide it between themselves in the ratio of their ages.
Clara is 8 and Di is 11.
How much does Di get?

15 78 CDs are given to Nasser and Mandy.
They divide them in the ratio 7 : 6.
Mandy gets the larger number.
How many does Nasser get?

16 The fuel tank on a moped holds 20 litres.
The fuel must be a mixture of oil and petrol in the ratio 1 : 50.
Work out how much oil is needed for a full tank.

17 Concentrated orange juice is diluted with water in the ratio 2 : 5.
How many millilitres of concentrated orange are needed to make up 3.5 litres of juice to drink?

18 The area of a garden is 357 square metres.
The ratio of the area of lawn to the area of flowerbed is 12 : 5.
   a Find the area of the lawn.
   b Work out the area of the flowerbed.

19 The tank on a chemical sprayer holds 5 litres.
    The instructions on a bottle of moss killer are

    'dilute in the ratio 1 : 300'.

    a   How much moss killer is needed to make a full tank of spray?
    b   Work out the quantity of moss killer needed to make 3 litres
        of spray.

20 Bronze contains copper and tin in the ratio 3 : 22 by weight.
    Find the weight of copper needed to make 5 kg of bronze.

21 A fruit drink is made by mixing orange juice, apple juice and
    mango juice in the ratio 4 : 3 : 2.
    Work out the quantity of mango juice in 1 litre of this drink.

22 Jo, Andy and Greg bought a house together.
    Jo contributed £50 000, Andy contributed £35 000 and Greg
    contributed £70 000.
    They sell the house for £250 000 and they divide this sum in the
    ratio of their contributions. How much does each person get?

ICT task

Use an Internet search engine to find the standard paper sizes used in
the UK.
Enter these sizes in a spreadsheet. Use the spreadsheet to investigate
the ratios of width to length, and the relationships between different
paper sizes. Write down anything you notice.

Make sure that what
you write makes
sense. Also make sure
that anyone else who
reads your work can
understand it.

Summary of key points

- Ratios can be simplified by dividing the parts by the same number.
- Units can be left out of a ratio as long as they are the same.
- Two quantities are proportional when they are in a constant ratio to each other. This means that if one quantity is multiplied by a number, the other quantity is multiplied by the same number.
- You can divide a quantity in the ratio $2:3$ by finding $\frac{2}{5}$ of the quantity and $\frac{3}{5}$ of the quantity.

Most students who get GRADE E or above can:
- use ratios in maps, scale drawings and gears
- understand and use proportion.

Most students who get GRADE C can also:
- divide quantities in a given ratio with three parts.

Glossary

Direct proportion	two quantities that are always in the same ratio
Division in a given ratio	dividing a quantity so that the parts are in a given ratio
Gear ratio	the ratio of the number of teeth on one wheel to the number of teeth on the other wheel
Map ratio	the ratio between lengths on the map and lengths on the ground
Ratio	a comparison between the sizes of quantities
Proportion	the same as direct proportion, i.e. two quantities that are always in the same ratio
Scale drawing	a drawing of an object that is proportional to the object
The symbol :	used in ratios to mean 'to', e.g. $2:3$ means '2 to 3'

10 Measures 2

This chapter will show you:
- ✓ the common units used for measuring time and temperature
- ✓ the relationships between different units for measuring a quantity
- ✓ compound measures such as speed
- ✓ how to convert between different currencies

Before you start you need to know:
- ✓ how to multiply and divide by 10, 100, 1000,...
- ✓ how to add, subtract, multiply and divide with decimals
- ✓ how to find one quantity as a fraction of another
- ✓ how to find a fraction of a quantity
- ✓ how to work with negative numbers

10.1 Time

Time is measured in millennia (1 millennium = 1000 years), centuries, decades, years, months, weeks, days, hours, minutes and seconds.

There are 12 months in a year but the number of days in each month varies.
There are 365 days in a year, except for leap years when there are 366.

The relationships between weeks, days, hours, minutes and seconds are:

$$1 \text{ week} = 7 \text{ days}$$
$$1 \text{ day} = 24 \text{ hours}$$
$$1 \text{ hour} = 60 \text{ minutes}$$
$$1 \text{ minute} = 60 \text{ seconds}$$

When you change units of time, remember that you multiply when you change to a smaller unit, and you divide when you change to a larger unit.

> Remember:
> Thirty days has September, April, June and November.
> All the rest have thirty-one except for February alone which has twenty-eight days clear and twenty-nine days each leap year.

> Leap years occur once every four years. 2008, 2012, 2016 are leap years.

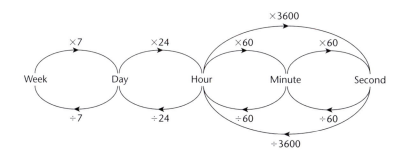

There are two ways of measuring the time of day.

The **24-hour clock** uses the full 24 hours in a day, measuring from midnight through to the next midnight.

The time is given as a four-figure number, for example, 1346 or 0730.
The first two figures give the hours and the second two figures give the minutes.
Sometimes you will see a space or a colon between the hours and the minutes: 13 46 or 13:46.

The **12-hour** clock uses the 12 hours from midnight to midday as a.m. times and the 12 hours from midday to midnight as p.m. times.

The time is written as a number of hours and a number of minutes followed by a.m. or p.m., for example 1.46 p.m. or 7.30 a.m.
The hours and the minutes are usually separated by a full stop or a colon.

12-hour clock: 24-hour clock:
2.56 pm   14.56

In the 24-hour clock, it is clear that 0000 means midnight and 1200 means midday. In the 12-hour clock, you need to write 'midnight' or 'midday' because 12.00 could mean either.

Example 1

Give the time on this clock in a.m. or p.m. time.

The time on this clock is 5.08 p.m.

24-hour times greater than 1200 are p.m. times. Subtract 12 to convert to 12-hour time.

Exam practice 10A

1   Look at this calendar.

Mon.	Tue.	Wed.	Thur.	Fri.	Sat.	Sun.
		1	2	3	4	5
6	7	8	9	10	11	12
13	14	15	16	17	18	19
20	21	22	23	24	25	26
27	28	29	30			

a   Which month is this – August, September or October?
b   What day of the week was on the 16th?
c   Today is the 21st of the month.
    i   What was the date a week ago today?
    ii   What is the date the day after tomorrow?

2   Yesterday was Wednesday 17th May.
    What was the date a week last Tuesday?

3   a   Steph leaves on 17th August for a holiday.
        She is away for ten nights.
        On what date does she return?

b George's school closes on 18th July.
   It starts again on 6th September.
   How many full days of summer holiday does George have?
c Sharon starts work on 1st September.
   She gets paid on the twentieth of each month.
   How many times does she get paid before Christmas?

4 The dates of birth of three children are:

   Joy 14-3-95   Donna 14-1-95   Rosemary 14-8-95

   a Who is the eldest?
   b Who is the youngest?
   c In which year will the youngest be 30?

5 The president of the local soccer club is elected every year at the
   Annual General Meeting.
   This is a list of the presidents since the club was formed.
   a For how many years was P. Leach president?
   b Who was president for the greatest number of years?
   c Assuming that C.D. Wilkins continues as president, which
      year will he be elected to begin his 25th year?

1909-23	S. Davidge
1923-29	S. James
1929-38	D.S. Short
1938-55	P. Leach
1955-63	N. Chicken
1963-77	D.S. Short
1977-80	T. Peters
1980-88	D.C. Bird
1988-	C.D. Wilkins

A person can be president more than once.

6 Write:
   a 170 minutes in hours and minutes
   b 490 hours in days and hours
   c 2 minutes and 6 seconds in seconds
   d $2\frac{1}{2}$ hours in minutes.

7 Find:
   a 40 minutes as a fraction of an hour
   b 36 seconds as a fraction of an hour.

8 These show the times at the beginning and
   end of Brett's morning geography lesson.
   a What time did the lesson start?
   b What time did the lesson end?
   c How long was the lesson?

Lesson begins   Lesson ends

9 My bus is due at 4.25 p.m.
   a How long should I have to wait?
   b Write 4.25 p.m. in 24-hour time.

10 Britt got home at twenty to four in the afternoon.
    She immediately turned on Channel 4.
    a How much of Countdown had she missed?
    b How long did she have to wait until The Simpsons started?

Channel 4	
3.30	Countdown
5.00	Richard & Judy
6.00	The Simpsons
6.30	Hollyoaks
7.00	Channel 4 News
7.55	Minute Wonder
8.00	Dispatches
9.00	Bodyshock
10.00	ER

11 Find the number of hours and minutes between
    a 8.30 a.m. and 10.15 a.m. the same day
    b 9.30 a.m. and 6.10 p.m. the same day
    c 10.20 p.m. and 12.30 a.m. the next day.

12 Mary sets the video recorder to start recording a programme at
4.50 p.m.
The programme lasts for $2\frac{1}{4}$ hours.
What time should she set the video to stop recording?

13 The time needed to cook a chicken is 40 minutes per kilogram
plus 40 minutes.
How long should it take to cook a $2\frac{1}{2}$ kg chicken?

14 Sue's bus is due at 2010.
    a How many minutes should she have to wait?
    b Write 2010 using the 12-hour clock.

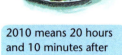
2010 means 20 hours and 10 minutes after midnight.

15 One day last winter, street lights were switched on at 1627 and
switched off at 0725 the next day.
How long were the street lights on for?

16 Find the period of time between
    a 0320 hours and 0950 hours on the same day
    b 0535 hours and 1404 hours on the same day
    c 2100 hours and 0500 hours next day
    d 0000 and 0305 the same day.

Remember 0320 means 3 hours and 20 minutes after midnight and 0305 means 3 hours and 5 minutes after midnight.

17 A train leaves King's Cross for Edinburgh.
The journey should take 4 hours 10 minutes.
The train leaves King's Cross on time at 1440 hours and is
25 minutes late arriving in Edinburgh.
Work out the time the train arrives in Edinburgh.

18 A flight from Cardiff to Prague takes 2 hours 10 minutes.
The plane leaves Cardiff at 1655.
Find the time the plane arrives.

19 The bus service from Dixford to Limpley runs twice a day.
Here is the timetable.

		No.81	No. 83
Dixford		0945	1420
Farm Hill		1004	1439
Bewley	*arr*	1056	1531
	dep	1116	1545
Peters Farm		1129	1559
Limpley		1207	1637

    a How long does each bus take to go from Dixford to Limpley?
    b Which two bus stops do you think are closest together?
    Give a reason for your answer.

A01

20 Here is part of a train timetable on the Taunton-Plymouth route.

Taunton	d	1757	1833	1843	1849	1904	1927	—	—
Tiverton Park	d	—	—	1857	—	1918	—	—	—
Exeter St. David's	d	1837	1901	1916	1922	1935	1956	—	2010
Exeter St. Thomas	d	—	—	—	—	—	—	—	2013
Starcross	d	—	—	—	—	—	—	—	2022
Dawlish Warren	d	—	—	—	—	—	—	—	2027
Dawlish	d	—	1914	—	—	—	—	—	2031
Teignmouth	d	—	1919	—	—	—	—	—	2036
Newton Abbott	a	—	1926	1935	1943	1954	2015	—	2043
Newton Abbott	d	—	1928	1937	1945	1956	2017	2025	2045
Torre	d	—	—	—	—	—	—	2033	—
Torquay	d	—	1942	—	—	—	—	2036	—
Paignton	a	—	1948	—	—	—	—	2041	—
Totnes	d	—	—	1951	—	—	—	—	2058 a
Plymouth	a	1943	—	2021	2029	2036	2057	—	—

d means depart.
a means arrive.

a Sara leaves Taunton at 1849 to visit a friend who lives in Newton Abbot.
What time will she get to Newton Abbot?

b Ian misses the 1928 at Newton Abbot by 10 minutes.
How long must he wait for the next train to Paignton?

21 This is part of a train timetable from Manchester Piccadilly to London Euston.

Manchester Piccadilly	1130	1133	1230	——	1330	1333	1430	1433	1530	——
Stockport	1138	1142	1238	——	1338	1342	1438	1442	1538	——
Wilmslow	——	1150	——	1250	——	1350	——	1450	——	1527
Crewe	——	1225	——	1325	——	1425	——	1525	——	1608
Macclesfield	1152	——	1252	——	1352	——	1452	——	1552	——
Stoke-on-Trent	1212	——	1312	——	1412	——	1512	——	1612	——
Rugby	——	1325	1359	——	——	1527	——	1628	——	——
Milton Keynes Central	——	——	1424	1447	——	——	1618	1653	——	1725
Watford Junction	1342	1412	1513	1529	1542	1614	1659	——	1742	1753
London Euston	1405	1435	1510	1534	1605	1637	1705	1739	1805	1816

Heavy type means 'through train'.
Light type means 'connecting train' e.g. the 1333 from Manchester to London means 'change at Crewe'.

a Greg catches the 1530 train from Manchester and it arrives in London on time. How long is his train journey?

b Angie has to be in London by 6 p.m.
She is travelling from Stoke-on-Trent.
What is the latest train she can catch?

22 James is paid £6.45 an hour.
He works a shift of 5 hours and 20 minutes.
How much does James earn for this shift?

23 An Internet café charges £1.30 an hour for the use of a computer.
Uheme spends £4.
How long can he use a computer for?
Give your answer in hours and minutes.

24 Janet has a pay-as-you-go mobile phone.
She buys a £5 top up card.
Off-peak calls cost 12p a minute.
How long can Janet talk at the off-peak rate with this top up card?
Give your answer in minutes and seconds.

25 a What time is shown on this parking meter?
    b Norma puts £5 into the meter.
       What is the latest time she can leave her parking space?

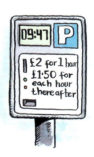

26 This parking meter shows the time when Clive
    arrives to park his car. The charge is £2 per hour
    or part of an hour. Clive wants to park his car
    until 4.35.
    a How much should he put into the meter?
    b He is 10 minutes late returning to his car.
       Has his parking meter run out? Give a reason
       for your answer.

10.2 Temperature

In Europe, temperature is usually measured in degrees Celsius.
The freezing point of water is zero degrees Celsius. This is written 0°C.
The boiling point of water is 100 degrees Celsius. This is written 100°C.

In the USA, temperature is usually measured in degrees Fahrenheit.
The freezing point of water is 32 degrees Fahrenheit. This is written 32°F.
The boiling point of water is 212°F.

You can convert between these two temperature scales.

To convert from Fahrenheit to Celsius, use this flow chart.

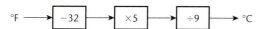

To convert from Celsius to Fahrenheit, use this flow chart.

Example 2

Convert   a 64°F to degrees Celsius
          b 85°C to degrees Fahrenheit.

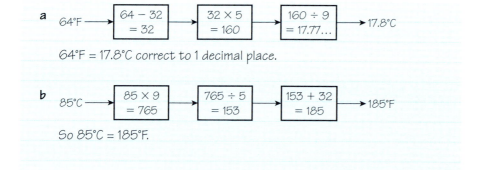

64°F = 17.8°C correct to 1 decimal place.

So 85°C = 185°F.

Exam practice 10B

1 **a** What temperature is shown on this thermometer?

b What temperature is shown on this thermometer?

Make sure you use the right units.

You do not have to convert between Celsius and Fahrenheit to answer this.

A01

c Which thermometer shows the higher temperature? Give a reason for your answer.

2 This thermometer is marked in degrees Fahrenheit and degrees Celsius.

a What Fahrenheit temperature does the thermometer show?

b What Celsius temperature does the thermometer show?

The temperature goes down by 20°C.

c What is the new Celsius reading?

d What is the new Fahrenheit reading?

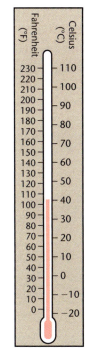

3 Use the flow charts on page 138 to convert

a 30°C to degrees Fahrenheit **b** −10°C to degrees Fahrenheit

c 80°F to degrees Celsius **d** 10°F to degrees Celsius.

4 In August 2003, the temperature in London reached a record high of 101°F. How many degrees Celsius is 101°F?

Use the flowchart on page 138.

5 The formula for converting a temperature of C degrees Celsius to F degrees Fahrenheit can be written as $F = \dfrac{9 \times C}{5} + 32$.

Use this formula to convert 45°C to degrees Fahrenheit.

You need to write 45 instead of C in the formula.

6 The formula for converting a temperature of F degrees Fahrenheit to C degrees Celsius can be written as $C = \dfrac{5(F - 32)}{9}$.

Use this formula to convert 250°F to degrees Celsius.

10.3 Speed

When something moves it covers distance.
Speed measures the distance covered per unit of time.

Speed is a **compound measure** because it combines distance and time.

$$\text{average speed} = \frac{\text{distance covered}}{\text{time taken}}$$

The most common metric units of speed are kilometres per hour (km/h) and metres per second (m/s).

The only Imperial unit of speed in everyday use is miles per hour (mph).

> A car travels 80 miles in two hours. The car travels 40 miles each hour so its speed is 40 mph (mph is short for miles per hour) .

> When you travel, your speed is likely to vary from time to time. You can give a speed for a whole journey by working out the average speed. You do this by finding the distance covered then dividing it by the time taken.

Example 3

Pedro walks 5 km in $1\frac{1}{2}$ hours.
a Find his average speed.
b Pedro stops for 10 minutes then walks a further 3 km in 50 minutes.
  Find Pedro's average speed for the whole of his journey.

a   Pedro's average speed is $5 \div 1\frac{1}{2}$ km/h

                $= 3.3$ km/h correct to one decimal place.

b   Total distance is 5 km + 3 km = 8 km

     Total time is $1\frac{1}{2}$ hours + 10 min + 50 min

            $= 1\frac{1}{2}$ h + 60 min = $2\frac{1}{2}$ h

     Average speed = $8 \div 2\frac{1}{2}$ km/h = 3.2 km/h.

> To find his average speed you need the total distance and the total time.

> To find the speed in kilometres per hour, you need to write the time in hours.

You can convert a speed given in one unit to another unit.

Example 4

Change a speed of 10 km/h to a speed in mph.

        8 km ≈ 5 miles

so   1 km ≈ 5 ÷ 8 miles = 0.625 miles

and   10 km ≈ 10 × 0.625 miles

           = 6.25 miles.

So   10 km/h ≈ 6.3 mph correct to 1 decimal place.

> 10 km/h means 10 km every hour.
> To convert 10km/h to mph, you need to convert 10 km to a number of miles.
> Use 8 km ≈ 5 miles.

Exam practice 10C

Some question have exact answers. When the answer is not exact, you need to decide how accurate your answer should be. In general, 1 decimal place is accurate enough. Remember to put units in your answers.

1   Robert cycled 30 km in 2 hours.
    Find his average speed.

2   A train travelled 450 miles in 4 hours.
    Work out the average speed of the train.

3   Keith took 30 minutes to walk 2 kilometres.
    What was his average speed in kilometres per hour?

4   Greg drove 75 miles in $1\frac{1}{2}$ hours.
    Work out his average speed in miles per hour.

5   This chart shows the distance between six cities.

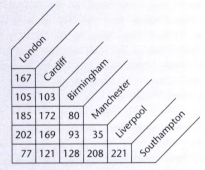

Distances are in miles

A car travels from Manchester to Southampton in 4 hours.
What is its average speed?

6   Bejay cycled 4.5 km in 10 minutes.
    Find his average speed in kilometres per hour.

To find a speed in kilometres per hour, you need the distance in kilometres and the time in hours.

7   A bus took 30 minutes to travel 3 miles.
    What was the average speed of the bus?
    Give your answer in miles per hour.

To find a speed in miles per hour, you need the distance in miles and the time in hours.

8   a   Vicki took 20 minutes to walk $1\frac{1}{4}$ miles.
        What was her average speed in miles per hour?
    b   Rupert drove 50 miles in 75 minutes.
        Find his average speed in miles per hour.

9   Fred shot an arrow at a target 90 metres away.
    The arrow was in flight for 2 seconds.
    Work out the average speed of the arrow in metres per second.

10 Mohammed completed a half-marathon of 13 miles.
    He ran for 2 hours and walked for 30 minutes.
    Find Mohammed's average speed in miles per hour.

11 Sam used his bicycle to travel 15 miles to work.
    He cycled for 30 minutes, pushed his bicycle for 5 minutes and
    cycled the rest of the distance in 10 minutes.
    Find Sam's average speed in miles per hour.

12 The average speed of a car in the rush hour is 4 mph.
    How long does the car take to travel 6 miles at this speed?

You can find the time by dividing the distance by the speed.

13 Ann travels between London Waterloo and Brussels Midi.

LONDON	07:39	07:43	08:12	08:39	09:09	10:39	10:42	11:39	12:09	12:39	14:09
Ashford	08:29	—	—	09:30	09:59	—	—	—	12:59	13:30	14:59
Calais	—	—	—	10:56	—	—	—	—	14:31	—	—
Lille	—	—	—	11:29	—	—	13:24	14:21	—	15:29	—
BRUSSELS	—	11:03	—	12:10	—	—	14:05	—	—	16:10	—
PARIS	11:23		11:47		12:53	14:17		15:23	15:59		17:53

She catches the 1042 train from London Waterloo.
The distance between London and Brussels is 370 km.
a Work out the average speed of the train assuming that it arrives on time.
b The train arrives 30 minutes late. What is the average speed?

14 The speed limit on motorways is 70 mph.
Use 1 mile = 1.61 km to give this speed limit in km/h.

15 a Write down the speed shown on this dial.   **b** Write down the speed on this dial.

A01

c Which dial shows the faster speed?
Give a reason for your answer.

10.4 Exchange rates

Each country has its own units of money.
The units of money are called **currencies**.

In the United Kingdom, the currency is pounds sterling and the symbol is £ where £1 = 100 pence.
In parts of Europe, the currency is the Euro and the symbol is € where €1 = 100 cents.

Did you know

that several countries use the dollar as the name of their currency but they are not all worth the same amount?
The symbol for the dollar is $ where $1 = 100 cents.

The dollar symbol often has letters with it to distinguish between the different currencies. For example, US$ in the USA and NZ$ in New Zealand.

You can convert one currency into another. To do this you need to know the **exchange rate**.
The exchange rate tells you what one unit of one currency is equal to in another currency.

The exchange rates between most currencies change from day to day. This is because currencies can be bought and sold in the same way as, for example, houses or coffee.

Example 5

Use the exchange rate £1 = US$1.75 to convert:

a £2.50 into US dollars. (Give your answer to the nearest cent.)

b US$25 into pounds sterling. (Give your answer to the nearest penny.)

a £2.50 = US$ 2.50 × 1.75

      = US$ 4.375

      = US$ 4.38 to the nearest cent

> Each £1 is worth 1.75 dollars so you convert pounds to dollars by multiplying by 1.75.

b US$25 = £25 ÷ 1.75

      = £14.285…

      = £14.29 to the nearest penny.

> To convert dollars into pounds you do the reverse; divide the number of dollars by 1.75.

Exam practice 10D

1 This list shows the exchange rate for some currencies.

> Give answers to the appropriate degree of accuracy.

      £1 = €1.69
      £1 = 2.68 Turkish Lira (l)
      £1 = 1.80 US dollars

  a Freda changed £50 into Euros.
    How many Euros did she get?

  b Pete changed £200 into Turkish lira.
    Work out how many lira he got.

  c Jon changed £150 into US dollars.
    Find how many dollars he got.

2 William changed £200 into US dollars.
  The exchange rate was £1 = $1.70
  How many dollars did he get?

3 Sara changed £600 into Australian dollars.
  The exchange rate was £1 = $2.30.
  How many dollars did she get?

4 Henry changed 500 Swiss francs into pounds.
  He got £1 for every 2.60 francs.
  How many pounds and pence did he get?

> You need to divide 500 by 2.60.

5 Jo paid a hotel bill for €500.
  The exchange rate was £1 = €1.58.
  Find the amount of the hotel bill in pounds.

6 Francis paid for a flight in Japan with his credit card.
  The amount on his statement was £500.
  What was the cost of the flight in Japanese yen if the exchange rate was 160 yen for £1?

7   The cost of a bottle of water in Germany was €1.08.
Find the cost in pounds sterling.
The exchange rate was £1 = €1.20.

8   David spends 50 US dollars on a camera.
The exchange rate is US$1.85 for £1.
Find the cost of the camera in pounds.

9   Sally went to Australia.
The exchange rate was £1 = 2.60 Australian dollars.
   a   Sally bought a bowl costing $500.
      Find the price of the bowl in pounds.
   b   Sally came home with 40 Australian dollars and changed
      them back into pounds.
      How many pounds did she get?

10   George went to France.
He used £2 = €3 to get a rough idea of prices in pounds sterling.
   a   George estimated that a CD cost £12.
      What was the price in Euros?
   b   A pair of designer jeans was €120.
      What did George estimate the price to be in pounds sterling?

11   Sonja changed £900 into Indian rupees. She got 55 440 rupees.
What was the exchange rate?

12   This is a conversion graph.

> Find £26 on the axis
> labelled Pounds (£).
> Then go up to the line.
> Now go across to the
> axis labelled Euros (€).

Use it to convert
   a   £26 into Euros
   b   €20 into pounds.
Joe paid €35 for a meal.
   c   What was the price of the meal in pounds?

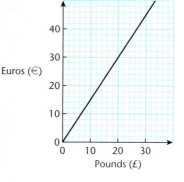

13   A bottle of water cost 48p in London.
The same bottle of water cost €0.70 in Madrid when the
exchange rate was €1.25 to the pound.
In which city was the bottle of water cheaper?

14   Dennis went to Paris
   a   Dennis changed £100 into Euros.
      He got €152.
      What was the exchange rate?
   b   When Dennis got back to London, he changed €30 into
      pounds.
      He got £18.29.
      What was the exchange rate for this transaction?

15 Penny is going to Florida.
She went to her bank to change some money into US dollars.
She saw this notice.

BANK BUYS	BANK SELLS
£1 US$ 1.90	US$ 1.70

> When you change £ to $ the bank is selling you dollars.

a Penny changes £75 into US dollars.
How many does she get?
b When she gets back, how much will she get if she changes US$40 into sterling?

16 Pablo went to Spain.
He paid €250 for a pair of designer sunglasses.
Pierre went to Florida.
He paid US$280 for a pair of the same designer sunglasses.
The same sunglasses were on sale in London for £160.
Use £1 = $1.85 and £1 = €1.49 to find where these sunglasses were cheapest.

Summary of key points

- Temperature is measured in degrees Celsius or in degrees Fahrenheit.
- The 24-hour clock measures the time of day from midnight and is written as a four figure number, e.g. 0730 means 7 hours and 30 minutes after midnight and 1930 means 7 hours and 30 minutes after midday.
- The 12-hour clock measures the time of day from midnight to midday as a.m. times and from midday to midnight as p.m. times.
- Speed is measured in kilometres per hour (km/h) or miles per hour (mph).
- average speed $= \dfrac{\text{total distance}}{\text{total time}}$
- You can change an amount in pounds to an amount in another currency by multiplying by the exchange rate.
- You can change an amount in another currency back to pounds by dividing by the exchange rate.

Most students who get GRADE E or above can:
- use a timetable
- convert from one currency to another.

Most students who get GRADE C can also:
- work out average speeds.

Glossary

Average speed	total journey distance divided by total journey time
Compound measure	a unit that involves two or more simple units
Conversion graph	a graph that can be used to convert from one unit to another unit
Exchange rate	how much of another currency you will get for £1
Speed	the distance covered in one unit of time

Examination practice paper

1 Two shops are selling the same bars of chocolate.

Lo-prices 40 pence a bar	**Costsavers** Buy 2 bars for 90 pence

Ann wants to buy 2 bars of chocolate.
Explain why it is cheaper for her to buy the bars at Lo-prices.   *(2 marks)*

2 (a) Here is a list of numbers

         17   3   28   9   24   41

Write a number from the list that is
(i)   an odd number   *(1 mark)*
(ii)   a square number   *(1 mark)*
(iii)   a multiple of 6.   *(1 mark)*

(b) Write down all the factors of 21.   *(2 marks)*

3 The prices to visit a zoo are shown.

	Adult	Child
Monday to Friday	£16	£12.50
Saturday to Sunday	£17.50	£15

(a) Mrs Jones and her 2 children visit the zoo on Monday.
    How much do they pay in total?   *(2 marks)*

(b) Mr Khan buys 2 adult tickets and some child tickets for a Saturday visit.
    He pays a total of £125.

    How many child tickets did he buy?
    Show your working.   *(3 marks)*

4 The number of pupils in each Year group at a school is shown.

Year	7	8	9	10	11
Number of pupils	120	105	112	99	103

   (a) Work out the total number of pupils in the school.   *(1 mark)*

   (b) $\frac{2}{5}$ of the pupils in Year 7 are girls.
       Work out how many girls are in Year 7.   *(2 marks)*

   (c) A teacher says that half of the pupils in Year 11 are boys.
       Explain why the teacher is wrong.   *(1 mark)*

5 Use your calculator to work out

   (a) the square of 1.7   *(1 mark)*

   (b) the cube root of 343   *(1 mark)*

   (c) $\dfrac{6.2 \times 2.8}{5.4 + 1.9}$

       (i) Write down all the digits from your calculator display.   *(1 mark)*
       (ii) Write your answer to 2 decimal places.   *(1 mark)*

6 Here is part of a list of programme times for a television channel.

       News   13 00
       Weather   13 25
       Film   13 30
       Cartoons   15 25
       Soap   16 05
       Music Videos 16 35
       News   17 25

   (a) How many minutes are the Music Videos on for?   *(1 mark)*

   (b) Bill records the film on a blank DVD that has 4 hours of recording time.
       How much of the DVD is still available to record onto?
       Give your answer in hours and minutes.   *(3 marks)*

7 (a) A shop has a CD priced at £10.20 before a sale.
       In a sale the price is decreased by 15%.
       Work out the price of the CD in the sale.   *(3 marks)*

   (b) The number of these CDs sold by the shop increased from 150 per week to
       195 per week during the sale.
       Work out the percentage increase in the number of CDs sold per week. *(3 marks)*

8 The number of people at a football match is 24 600 to the nearest hundred.
   Write down

   (a) the least possible number of people at the match   *(1 mark)*

   (b) the greatest possible number of people at the match.   *(1 mark)*

Section B   Time allowed: 40 minutes   Calculators are not allowed

1 (a) Write $\frac{1}{2}$ as a decimal.   *(1 mark)*

   (b) Write 0.2 as a fraction.   *(1 mark)*

   (c) Write 60% as a decimal.   *(1 mark)*

2 Mary buys 4 cakes costing 60 pence each.
She pays with a £5 note.
How much change should she receive?   *(2 marks)*

3 Put these numbers in order of size.
Start with the smallest number.

   (a) 1001   1100   1010   *(1 mark)*

   (b) 0.6   0.075   0.39   *(1 mark)*

   (c) −2   1   −3   *(1 mark)*

4 Work out

   (a) 350 − 136   *(2 marks)*

   (b) 136 ÷ 4   *(1 mark)*

   (c) 40% of 90   *(2 marks)*

   (d) −4 × −3   *(1 mark)*

   (e) 236 × 47   *(3 marks)*

5 Paul is multiplying decimal numbers by 10

   $3.2 \times 10 = 32$ $5.8 \times 10 = 58$ $24.9 \times 10 = 249$ $0.6 \times 10 = 6$

He says that when you multiply a decimal number by 10 you always get a whole number.

Give an example that shows that Paul is not correct.   *(2 marks)*

6 Work out

   (a) 0.2×0.4 *(1 mark)*

   (b) 2^3 *(1 mark)*

   (c) $\frac{1}{4} + \frac{3}{8}$ *(2 marks)*

7 A car travels 9 miles in a time of 15 minutes.
Calculate the average speed of the car.
Give your answer in miles per hour.   *(2 marks)*

8 Mel buys milk in 1 pint bottles.
He drinks $\frac{2}{3}$ of a pint of milk a day.

What is the least number of pints that he needs to buy for one week.   *(3 marks)*

9 To make Gorgeous Green paint, blue and yellow paint are mixed together in the following ratio.

$$\text{blue} : \text{yellow} = 5 : 2$$

How many litres of blue paint will be needed to make 56 litres of Gorgeous Green paint?   *(2 marks)*

10 In 2004 there were 400 members of a health club.
The number of members has increased by 10% each year since 2004.

Work out the number of members in 2006.   *(2 marks)*

Answers

Exam practice 1A

1 a 63   b 49   c 707   d 327
  e 819   f 8008   g 6067   h 15 234
2 a fifty-six   b seventy-nine
  c four hundred and nine
  d one hundred and eighty-seven
  e seven hundred and thirty-four
  f three hundred and thirty
  g four hundred and twenty-six
  h nine thousand four hundred and eighty-
     eight
  i six thousand five hundred and ninety-three
  j seven thousand and sixty-five
3 a i 5 ii 6   b i 9 ii 0 iii 8
4 a 43, 55, 57, 61   b 27, 31, 49, 83
  c 77, 104, 293, 308   d 506, 560, 605, 650
  e 98, 845, 1088, 8876
  f 2033, 2303, 3032, 3302
5 a tens   b hundreds   c units
  d thousands   e ten thousands
6 a hundreds   b units   c thousands
  d ten thousands
7 a 974   b 479
8 a 540   b 405
9 a 2379   b 6532   c 5017   d 5864
10 a 2440   b 2040
11 543, 534, 453, 435, 354, 345

Exam practice 1B

1 a 10   b 21   c 26   d 18   e 20
  f 20   g 21   h 16   i 30   j 38
  k 25   l 28
2 a 21   b 33   c 24
  d 25   e 32   f 24
3 34, 41, 48, 55
4 38, 46, 54, 62
5 34 m

Exam practice 1C

1 a 79   b 97   c 65   d 308   e 559
2 a 19 and 31   b 31 and 69
3 a 47 and 53   b 31 and 89 or 41 and 79
4 No, 143 + 426 = 569.
5 a 399   b 779   c 1207 d 985   e 3302
  f 549   g 13 257 h 8146 i 1584 j 12 540
6 a 112   b 158   c 183   d 177   e 789
  f 998   g 851   h 1124 i 10 679 j 2991
7 843

Exam practice 1D

8 47 min
9 4957
10 a 31   b 77   c 67
   d 50   e 56   f 103

1 a 11   b 12   c 14   d 5   e 7   f 13
2 a 15   b 3   c 5   d 9   e 11   f 6
  g 381   h 336   i 52   j 1102 k 70   l 4
  m 236   n 3365   p 4722
3 301
4 89
5 a 14 take away 6 gives an 8.
  b Yes, 77 + 199 = 276
6 36
7 6483
8 240
9 76, 68, 60, 52, 44
10 159
11 a

2	7	6
9	5	1
4	3	8

   b

4	9	2
3	5	7
8	1	6

12 a 7   b 3   c 6   d 3
   e 7   f 5   g 2   h 8

Exam practice 1E

1 a 7   b 8   c 2   d 3
  e 2   f 13   g 0   h 0
2 36
3 Subtracts 7 rather than adds 7.
4 Subtracts 7 rather than adds 7.
5 a 230 b 80   c 158 d 35   e 19   f 181
6 a C   b B

Exam practice 1F

1 318
2 12 kg
3 44
4 1531
5 a 24, 19, 28, 23   b 11, 15, 19
6 560 cm
7 9
8 a 1890
  b 115
9 430 girls, 370 boys

Exam practice 1G

1 32
2 a 63   b 17
3 40
4 a 304   b 415   c 1899
   d 1632   e 852   f 2135
5 3612 g
6 192, 768
7 7, 14, 28, 56, 112
8 a 7   b 3   c 2

Exam practice 1H

1 a 450   b 6300   c 170
   d 26 000   e 5600   f 15 000
   g 370 000   h 25 000   i 84 000
2 a 460   b 6400   c 1290   d 2850
   e 9600   f 4860   g 45 600   h 64 800
   i 414 000   j 390 000   k 49 500   l 199 600
3 32 000
4 A and C, should be 6 digits in answer.
5 A and B. Answer has 5 digits and must begin
   with 90 (15 × 6).
6 8000
7 She forgot the 0 with the 50.

Exam practice 1I

1 a 672   b 1290   c 1428   d 2782
   e 3286   f 559   g 567   h 1558
   i 7844   j 7028   k 1564   l 5795
2 2332
3 713
4 2592
5 528
6 300
7 273

Exam practice 1J

1 a 29   b 21   c 103   d 294
   e 197   f 61   g 121   h 44
   i 1471   j 354   k 353   l 127
2 25
3 24
4 78
5 Yes, 376 ÷ 47 = 8.
6 27

Exam practice 1K

1 a 18, r3   b 12, r1   c 121, r5   d 89, r3
   e 182, r3   f 62, r7   g 22, r1   h 110, r2
   i 76, r2   j 77, r1   k 66, r3   l 111, r6
2 a 27   b 4
3 a 75   b 3
4 a 67   b 3
5 12, with 7 runners each time.
6 a 15   b 1
7 a 250 kg   b 62, 2 kg left over.
8 a 14   b 4
9 145
10 107

Exam practice 1L

1 a 25, r6   b 8, r7   c 9, r42
   d 30, r77   e 8, r76   f 4, r910
   g 278, r1   h 85, r12 i 7, r230
2 36
3 240
4 a 24   b 40
5 25
6 25
7 3000
8 £50 000
9 11

Exam practice 1M

1 a 12, r14   b 35, r0   c 313, r2   d 200, r13
   e 52, r9   f 16, r13 g 236, r0   h 51, r13
2 22
3 a 11   b 17
4 54
5 a 16   b 5
6 a 21   b 10
7 a 43   b 9

Exam practice 1N

1 a 16   b 16   c 4   d 1
   e 34   f 1   g 17   h 5
2 a 19   b 12   c 21   d 0   e 11
   f 1   g 5   h 15   i 9   j 15

Exam practice 1P

1 a 20   b 180   c 9   d 6   e 120   f 2
2 a 25   b 80   c 1   d 2   e 40   f 5
3 a 40   b 34

Exam practice 2A

1 a 3   b 0   c −1
2 a −9   b −4   c −2
3 a −11°C   b 7°C
4 a −10°C   b 4°C
5 a −11°C, −10°C, −8°C, −5°C, −3°C, 3°C,
      5°C, 9°C
   b −10°C, −8°C, −7°C, −4°C, −3°C,
      3°C, 7°C
6 a 5°C, 4°C, −6°C, −7°C, −9°C, −10°C, −12°C
   b 6°C, 5°C, −4°C, −5°C, −8°C, −11°C, −13°C
7 a Sydney   b Moscow
8 a −23, −7, −6, −4, 3, 15
   b 12, 9, −5, −8, −10, −13
9 a
   b

Exam practice 2B

1 a −4   b −7   c −10   d −4   e −1   f 4
2 a 1   b −4   c 6   d 5   e −3   f −5

3 a 9　b −2　c 13　d 4　e −4　f −20
4 a −2　b −1　c −2　d −15
5 a 2　b −3　c 4　d −5　e 3　f −6
6 5
7 −10°C
8 a 1°C　　　　b −7°C
9 a 5　　　　　b 11
10 83 degrees
11 a −1, 2　　　b −1, −4, −7
12 a 1 a.m.　　b 3 p.m.
13 a 5 a.m.　　b 3 p.m.　　c 9 a.m.
14 a 9 a.m.　　b 7 p.m.　　c 2 a.m.
15 10 p.m.
16 2 a.m. or 0200 on 5th May (2nd time!)

Exam practice 2C

1 a −6　b 20　c 20　d 12　e −15
　f −21　g −42　h −50　i −24
2 a −4　b −3　c −6　d −4　e −4
　f −5　g −2　h −5　i −5
3 a −8　b −26　c 7　d 10
4 a 28　b 23　c 1　d 2
　e −6　f 1
5 a 1　b 12　c −2
6 a i 5　ii 2　iii 4
　b i −3　ii −4　iii −5
7 a −7　b −4　c 4　d 3
8 −27, 81, −243
9 a −2　b 128, −256
10 −£500, i.e. £500 in debt
11 28
12 a £800　b £3800
13 a £851　b £206
14 a 60　b 10　c 30　d 6

Exam practice 3A

1 20, 46, 92
2 a 15　　　　b 72
3 a 7, 19, 31, 41　b 29, 11
4 a 3 + 3　b 3 + 7 or 5 + 5　c 7 + 7 or 3 + 11
5 a 3 + 3 + 5　b 5 + 5 + 5 or 3 + 5 + 7
　c 5 + 5 + 7 or 3 + 3 + 11 or 3 + 7 + 7
6 a 2, 3, 5, 7, 11, 13, 17, 19, 23, 29, 31, 37, 41, 43, 47
　b Yes, 7 and 43, 13 and 37, 19 and 31
7 4, 16, 36
8 Yes, 100 = 10 × 10.
9 No, no whole number × itself gives 56.
10 a 9, 11　b 2, 11　c 9
11 a 2　b 4 ÷ 2 = 2 which is even.

Exam practice 3B

1 a 1 × 8, 2 × 4
　b 1 × 12, 2 × 6, 3 × 4
　c 1 × 6, 2 × 3
　d 1 × 16, 2 × 8, 4 × 4
　e 1 × 18, 2 × 9, 3 × 6
　f 1 × 24, 2 × 12, 3 × 8, 4 × 6
2 Yes, 7 + 8 = 15 which is ÷ by 3, 4 + 8 = 12 which is ÷ by 3.

3 a yes　b yes　c yes
　d no　e yes　f yes
When the answer is yes, it is because the sum of digits can be divided exactly by 3.
4 Yes, 2 + 0 + 7 = 9 and 5 + 7 + 6 = 18. Both are ÷ by 9.
5 All except 725.
6 Any two from 1, 3, 5, 15.
7 The possible factors are
　a 1, 2, 3, 4, 6, 8, 12, 24
　b 1, 2, 13, 26
　c 1, 2, 7, 14.
8 a 2 and 3　b i 3 and 7　ii 2, 3, 7
9 a 1 , 2, 3, 6, 9, 18
　b 1, 2, 4, 7, 14, 28
　c 1, 2, 19, 38
10 a He's forgotten 20.
　b He's forgotten 10.
11 a 40　b 31　c 54
12 a True, only even prime number is 2.
　b False, 9 is odd but not prime.
　c True, 7 + 8 + 9 + 3 = 27 which is ÷ by 9.

Exam practice 3C

1 a e.g. 14　b e.g. 18, 27　c e.g. 24, 36, 48
2 a Yes, 5 × 4 = 20.
　b No, 54 ÷ 7 is not exact.
　c No, 46 ÷ 6 is not exact.
　d Yes, 7 × 8 = 56.
3 a 28 or 42　b 36 and 72　c 28
4 Yes, 9 × 12 = 108.
5 Yes, 2 + 3 + 4 = 9 which is ÷ by 3.
6 Yes, 7 × 18 = 126.

Exam practice 3D

1 1 and 7
2 3
3 a No, 38 does not ÷ exactly by 9.
　b Yes, 32 = 4 × 8 and 48 = 6 × 8.
4 a 3　b 8　c 12
5 a 25　b 11　c 21　d 13　e 5　f 4
6 a 4　b 40
7 120
8 8 will not divide exactly into 36, 44 or 52.
9 a 50 cm　　　b 63
10 a 6 cm side squares　b 35
11 21
12 a 110 cm　　　b 15
13 3 m
14 a 36 cm　　　b 154

Exam practice 3E

1 a 18　b 28　c 16　d 24
2 a e.g. 60　b e.g. 60
3 a Yes, 140 = 2 × 2 × 5 × 7.
　b Yes, 126 = 2 × 7 × 9.
4 a 72　b 80
5 a 6　b 30　c 12　d 6　e 30
　f 36　g 72　h 60　i 90
6 a 36　　　b 48

7 No, both divide exactly into 72.

8 12 m

9 a £100   b £10

10 20 cm

11 60

12 45 cm

13 a 17   b 7   c 4 or 36   d 8 or 40   e 3

Exam practice 3F

1 a 2^3 b 3^4 c 5^4 d 7^6

2 No, it is 3^6.

3 a 32   b 25   c 9   d 81

4 a 3^2 b 5^2

5 a No, $3^3 = 3 \times 3 \times 3 = 27$.
   b Yes, $2^4 = 2 \times 2 \times 2 \times 2 = 16$.

6 1, 4, 9, 16, 25, 36, 49, 64, 81, 100, 121, 144, 169, 196, 225

7 a 125   b 216   c 64   d 512

8 a $2^2 \times 7^2$ b $3^3 \times 2^2$ c $2^2 \times 3^2 \times 5^2$
   d $3^2 \times 5 \times 7^4$ e $5^2 \times 13^3$ f $3^3 \times 5^2 \times 7^2$

9 a 108   b 225   c 180   d 126

10 $2^3 \times 5^2 = 8 \times 25 = 200$.
   You cannot multiply 2 by 5 and/or add the indices.

11 a 64   b 81

12 $(3^2)^3 = 9^3 = 729$ $3^2 = 9$ not 6

13 a 100   b 1000   c 100 000   d 1 000 000

14 a 10^2 b 10^1 c 10^6 d 10^8

15 a 2^1 b 2^4 c 2^5

16 $2^2 - 4$

17 $a = 2, b = 3, a^3 + b^3 = 2^3 + 3^3 = 8 + 27 = 35$

18 a True   b False   c False

Exam practice 3G

1 a 4   b 5   c 10   d 9
   e 13   f 12   g 6   h 15

2 No, $7 \times 7 = 49$ so last figure in answer must be 9.

3 ±20

4 2

5 $900 = 9 \times 100$, 30

6 $484 = 4 \times 121$ so $\sqrt{284} = 22$

7 a 18   b 60   c 25

8 a 5   b −2   c 10   d −3

9 8×1000, 20

10 8×27, 6

11 a 7   b 132   c 21   d 4

Exam practice 3H

1 a 3^7 b 7^8 c 9^{10} d 2^9 e 4^{16} f 5^8

2 a Add the indices not multiply them.   b 10^7

3 a 4^3 b 2^2 c 5^1 d 3^3 e 2^2 f 5^7

4 a 12^9 b 7^4 c 3^5 d 4^4 e 5^1 f 3^3

5 2^6

6 3^{12}

7 Multiply indices not add them.

8 a 1   b 4   c 1   d 1

Exam practice 3I

1 a $2^2 \times 3^2$ b $2 \times 3 \times 13$ c $3 \times 5^2 \times 7$
   d $2^3 \times 3 \times 11$ e $2^2 \times 5^2$ f $3^2 \times 5 \times 7$
   g $2^4 \times 3 \times 5$ h $2^2 \times 3 \times 5^2$ i $2 \times 7 \times 11$

2 a 11^2 b $3^2 \times 7$ c $2^4 \times 7$
   d 3×37 e $2^2 \times 11$ f $3^2 \times 7^2$

3 a 2^2 b 2^3 c 7^2 d 2^5 e 3^2 f 2^6

4 a $2^3 \times 3$ b 2×3^2 c 6   d 72

5 a 6   b 1404

Exam practice 4A

1 a 1   b 3   c 2   d 5

2 a 5   b 4   c 6   d 12

3 a 7   b 12   c 7   d 8

4 a $\frac{1}{3}$ b $\frac{4}{5}$ c $\frac{5}{8}$ d $\frac{3}{4}$ e $\frac{7}{12}$ f $\frac{7}{10}$

5 a $\frac{3}{7}$ b $\frac{8}{11}$ c $\frac{9}{12}$ d $\frac{5}{20}$

6 a $\frac{5}{6}$ b $\frac{1}{6}$ c $\frac{1}{6}$ d $\frac{3}{5}$
   e $\frac{5}{9}$ f $\frac{3}{4}$ g $\frac{5}{7}$ h $\frac{8}{13}$

7 a   b
  c   d
  e

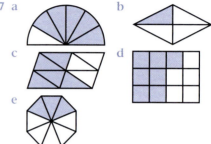

8 The lower two areas are much larger than the shaded area.

Exam practice 4B

1 a $\frac{4}{8}$ b $\frac{2}{8}$ c $\frac{6}{8}$

2 a $\frac{6}{12}$ b $\frac{8}{12}$ c $\frac{9}{12}$ d $\frac{2}{12}$

3 Missing numbers are   a 6   b 9   c 10.

4 a $\frac{6}{15}$ b $\frac{5}{15}$ c $\frac{10}{15}$

5 a $\frac{8}{24}$ b $\frac{12}{24}$ c $\frac{6}{24}$ d $\frac{18}{24}$ e $\frac{16}{24}$ f $\frac{3}{24}$

6 a $\frac{12}{32}$ b $\frac{2}{32}$ c $\frac{24}{32}$

7 a $\frac{6}{45}$ b $\frac{20}{45}$ c $\frac{30}{45}$

8 Missing numbers are a 4, 15, 10   b 6, 12, 15   c 2, 10, 30.

9 a $\frac{1}{4}$ b $\frac{1}{2}$ c $\frac{1}{2}$ d $\frac{3}{4}$ e $\frac{3}{4}$ f $\frac{2}{3}$
   g $\frac{2}{3}$ h $\frac{3}{4}$ i $\frac{3}{5}$ j $\frac{3}{4}$ k $\frac{4}{5}$ l $\frac{4}{5}$
   m $\frac{1}{4}$ n $\frac{2}{3}$ p $\frac{1}{3}$ q $\frac{1}{2}$ r $\frac{1}{2}$ s $\frac{3}{4}$

10 a $\frac{5}{6}$ b $\frac{1}{2}$ c $\frac{1}{4}$

11 **a** and **b**

Exam practice 4C

1 $\frac{15}{21}, \frac{14}{21}, \frac{5}{7}$

2 $\frac{10}{15}, \frac{12}{15}, \frac{2}{3}$

3 a $\frac{2}{9}$ b $\frac{2}{7}$ c $\frac{6}{7}$

4 $\frac{3}{4} = \frac{6}{8}$, yes $\frac{5}{8}$ is smaller than $\frac{6}{8}$.

Exam practice 4D

1 a $1\frac{1}{2}$ b $1\frac{1}{3}$ c $1\frac{2}{3}$ d $1\frac{3}{4}$ e $2\frac{2}{5}$ f $2\frac{1}{4}$
   g $3\frac{1}{3}$ h $3\frac{2}{3}$ i $3\frac{3}{4}$ j $2\frac{3}{4}$ k $2\frac{1}{3}$ l $2\frac{1}{5}$

2 a $\frac{5}{4}$ b $\frac{7}{3}$ c $\frac{13}{4}$ d $\frac{12}{5}$

  e $\frac{11}{4}$ f $\frac{7}{5}$ g $\frac{15}{4}$ h $\frac{17}{4}$

3 a $1\frac{1}{4}$ b $3\frac{1}{3}$ c $2\frac{2}{3}$ d $2\frac{2}{5}$ e $3\frac{1}{5}$ f $2\frac{4}{7}$

  g $1\frac{5}{7}$ h $1\frac{4}{5}$ i $1\frac{3}{7}$ j $1\frac{2}{9}$ k $1\frac{1}{11}$ l $1\frac{4}{7}$

11 $\frac{3}{16}$

12 $\frac{2}{3}$

13 $\frac{3}{8}$

14 $\frac{1}{15}$

Exam practice 4E

1 $\frac{1}{3}$

2 $\frac{2}{3}$

3 $\frac{1}{2}$

4 a $\frac{2}{9}$ b $\frac{7}{9}$

5 a $\frac{1}{4}$ b $\frac{1}{5}$ c $\frac{1}{10}$

6 $\frac{3}{8}$

7 a $\frac{1}{60}$ b $\frac{1}{6}$ c $\frac{3}{4}$ d $\frac{3}{5}$

8 $\frac{3}{5}$

9 $\frac{5}{6}$

10 $\frac{2}{3}$

11 $\frac{7}{16}$

12 $\frac{3}{7}$

Exam practice 4F

1 a $\frac{1}{2}$ b $\frac{3}{4}$ c $\frac{1}{2}$ d $\frac{2}{5}$ e 1   f $\frac{1}{6}$

2 $\frac{3}{4}$

3 a $\frac{7}{10}$ b $\frac{3}{8}$ c $\frac{1}{5}$

4 a $\frac{5}{8}$ b $\frac{11}{12}$ c $\frac{7}{8}$ d $\frac{3}{4}$

  e $\frac{1}{4}$ f $\frac{1}{8}$ g $\frac{1}{8}$ h $\frac{3}{8}$

5 a $\frac{13}{20}$ b $\frac{19}{20}$ c $\frac{1}{5}$ d $\frac{1}{6}$ e $\frac{8}{15}$

6 $\frac{5}{8}$

Exam practice 4G

1 a $3\frac{3}{4}$ b 5   c $6\frac{1}{2}$ d $1\frac{1}{4}$ e $2\frac{3}{8}$ f $1\frac{3}{8}$

2 a $3\frac{1}{20}$ b $4\frac{9}{20}$ c $4\frac{1}{4}$ d $\frac{7}{12}$ e $1\frac{5}{6}$ f $\frac{7}{12}$

3 $2\frac{1}{2}$ m

4 a $8\frac{1}{8}$ m   b $\frac{1}{2}$ m

5 $1\frac{3}{4}$ ft

Exam practice 4H

1 $\frac{2}{21}$

2 $\frac{12}{25}$

3 $\frac{5}{21}$

4 $\frac{3}{10}$

5 $\frac{2}{7}$

6 $\frac{1}{3}$

7 $\frac{1}{2}$

8 $\frac{4}{11}$

9 $\frac{1}{3}$

10 $\frac{1}{6}$

Exam practice 4I

1 a 1   b $1\frac{1}{8}$ c 5   d $\frac{1}{5}$

2 a $1\frac{2}{5}$ b 30   c $4\frac{1}{2}$ d 20

3 a $13\frac{1}{2}$ b 50   c 32   d 36

4 $\frac{15}{64}$ sq m

Exam practice 4J

1 $\frac{1}{6}$

2 2

3 8

4 $\frac{9}{5}$

5 $\frac{8}{5}$

6 $\frac{1}{20}$

7 $\frac{1}{100}$

8 $\frac{7}{12}$

9 $\frac{11}{5}$

10 $\frac{17}{20}$

Exam practice 4K

1 a 12   b 35

2 a 33   b 27   c 49   d 39   e $\frac{1}{40}$ f $\frac{1}{25}$

3 a $1\frac{1}{3}$ b 1   c $1\frac{1}{2}$ d $\frac{1}{6}$

4 a $\frac{7}{10}$ b 6   c $\frac{3}{8}$ d $\frac{27}{28}$

5 a $\frac{1}{6}$ b $1\frac{2}{3}$

6 a 20   b $\frac{3}{10}$ m

7 a $\frac{1}{40}$ b 1.5

Exam practice 4L

1 a   b

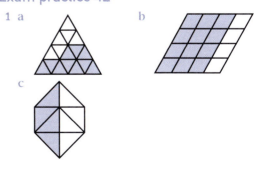

  c

2 a £6   b 4 days   c 8 hours   d 40 sec.

3 a £7   b 27 cm   c 3 m   d $22\frac{1}{2}$ t

4 a 32 euros b 219 days   c 34 kg   d 72p

5 a 33 miles b 21 litres c 28 km   d 21 hr

6 a 21p   b 28 cm

  c 292 days d 35 min

7 2
8 4
9 £16
10 £109
11 £300
12 £585
13 a £7   b £21
14 a £14   b £126
15 a 7000 kg   b 5000 kg
16 £64
17 a i 36 m   ii 15 m   b 9 m
18 a 9 km   b 6 km   c 3 km
19 a i 25   ii 24   b 11
20 Anne £2500, Julian £1000, Cheryl £1500

Exam practice 5A

1 a 4.3   b 8.76   c 7.044
   d 0.05   e 18.36   f 1.407
2 a three point eight eight
   b six point nought three
   c nought point nought six five
   d eleven point nought seven
3 a 7 hundredths   b 7 tens
   c 7 hundredths   d 7 tenths
   e 7 units   f 7 thousandths
4 a 4 hundredths   b 4 units
   c 4 tenths   d 4 hundredths
   e 4 thousandths   f 4 hundredths
5 a 0   b 3
6 a <   b >   c <
   d >   e <   f >
7 a 4.83, 6.29, 6.76   b 9.03, 9.07, 9.18, 9.51
8 a 14.09, 13.75, 12.6, 12.55
   b 7.555, 7.55, 7.5, 7.05
9 0.022, 0.202, 0.22, 2.2
10 a 0.4   b 0.15   c 0.37   d 0.2
11 0.555, 0.55, 0.505, 0.055

Exam practice 5B

1 a $\frac{1}{5}$ b $\frac{1}{2}$ c $\frac{3}{50}$ d $\frac{7}{10}$ e $\frac{4}{5}$ f $\frac{1}{40}$
2 a $\frac{1}{4}$ b $\frac{2}{25}$ c $1\frac{2}{5}$ d $\frac{1}{8}$ e $2\frac{3}{50}$ f $5\frac{1}{200}$
3 a 0.520   b 0.025
4 $\frac{17}{20}$
5 $\frac{17}{25}$
6 $\frac{12}{25}$

Exam practice 5C

1 a 1.9   b 2.35   c 1.8   d 1.38
   e 1.94   f 1.3   g 0.41   h 1.13
2 a 5.67   b 0.46   c 0.7   d 4.77
   e 10.5   f 12   g 3.7   h 4.3
3 2.15 kg
4 91.95 kg
5 95.5 kg
6 2.24 km
7 1.7 cm
8 18.08 cm

9 £ 2.85
10 a £6.09   b £ 3.91
11 a 18.8 cm   b 9.5 cm
12 18.78 mm
13 9.22 mm
14 3638.8 km

Exam practice 5D

1 a 250   b 0.66   c 24400   d 10
2 a 0.46   b 0.000 85   c 1.2   d 0.096
3 7
4 3.2

Exam practice 5E

1 a 1.6   b 3.2   c 7.2   d 6.5
   e 90   f 32   g 36   h 140
2 a 0.03   b 0.12   c 0.005   d 0.021
3 a 0.48   b 0.024   c 0.0088   d 0.0021
   e 0.12   f 0.01   g 0.45   h 0.0001
4 a 3   b 10   c 20   d 0.3
   e 20   f 30   g 40   h 2
5 a 3.24 kg   b 13.12 cm
   c 62 mm   d 3.78 m
6 0.045
7 a 990   b 333.75   c 7   d 0.9
   e 3.072   f 30   g 0.135   h 0.07
8 a 0.4   b 4.256   c 3   d 4.473
   e 640   f 2000   g 1.2   h 73.5
9 a 0.125   b 0.343   c 0.064
10 a £170.30   b £1.89
11 £2980.40   12 £2.21
13 19.68 mm   14 0.251 m
15 28   16 £4.09
17 95

Exam practice 5F

1 missing values are £26.40, £18.90, 22, £52.34
2 a missing values are £94.95, £16.66, 3, £115.63
   b £31.65
3 a missing values are £1.60, £1.36, 0.5, £4.97
   b £5.03
4 a missing values are £97.50, £37.25, 5, £142.50
   b £48.75
5 a missing values are £5.60, £0.73, £2.37, £0.5,
      £9.42
   b £6.57
6 a missing values are £5.40, £2.90, £10.65, 3,
      £21.92
   b £22.72

Exam practice 5G

1 a 0.2   b 0.125   c 0.75   d 0.6
2 a 0.8   b 0.375   c 0.15
   d 0.25   e 0.875   f 0.24
3 a 0.45   b 0.47
4 a 0.75   b $\frac{3}{4}$
5 $\frac{1}{8}$, 0.2, $\frac{1}{4}$, 0.35
6 a >   b >   c >   d >   e <   f <
7 $\frac{6}{25}$, $\frac{3}{16}$, $\frac{2}{13}$, 0.105, 0.05

8 a 5   b 1.25   c 0.625
   d 0.4   e 0.666...
9 a $0.\dot{6}$ b $0.\dot{1}1428 5\dot{7}$ c $0.1\dot{6}$ d $0.2\dot{6}$
   e $0.\dot{2}$ f $0.9\dot{0}$ g $0.08\dot{3}$ h $0.\dot{3}84615$
10 0.316316316
11 0.461538461538
12 a $0.1\dot{3}$ b $0.01\dot{3}$ c $0.001\dot{3}$

Exam practice 5H

1 a 5500   b 316 000   c 4 155 000
   d 577.8   e 13 000   f 9150
   g 80 220   h 200 400 000   i 7 400 000
2 1 000 000 000 000 decibels
3 150 000 000
4 9
5 a 48 000 000 000
   b 4.8 10, 4.8×10^{10} or equivalent
6 2020 000 000 000

Exam practice 5I

1 a 4π b 1.44π c 2.64π
   d 0.81π e 5.96π f 2.25π
2 3 is not an exact value for π.

Exam practice 6A

1 a 230   b 70   c 150
   d 630   e 40   f 160
2 a 300   b 1200   c 500
   d 3800   e 2000   f 7600
3 a 8   b 4   c 6
   d 2   e 19   f 1
4 a 7000   b 3000
   c 7000   d 16 000
   e 28 000   f 2000
5 a 6780   b 6800   c 7000
   d 96 000   e 95 700   f 95 680
6 a 200   b 160   c 160
   d 2180   e 2600   f 440
7 54 000
8 £25 900
9 59 000 000

Exam practice 6B

1 a 2.5   b 12.6   c 124.8
   d 24.2   e 3.2   f 0.1
2 a 6.57   b 0.35   c 12.66
   d 0.03   e 4.80   f 0.80
3 a 1.557   b 0.481   c 2.356
   d 0.015   e 0.104   f 0.004
4 a 0.69   b 40.38   c 28.8
   d 13.5   e 1.00   f 77.998
5 a 2.253   b 1.0028   c 0.05
   d 300   e 40   f 139.8
6 a 910   b 0.48   c 400
7 14.9
8 1.267 sec

Exam practice 6C

1 46 kg

2 0.009 cm
3 a 42p   b 13p
4 a £15   b £2.14 (nearest p)
5 a 0.33   b 0.14
6 a 0.67   b 0.11   c 0.17
   d 0.27   e 0.22   f 0.09
   g 0.08   h 0.38 (all to 2 d.p.)
7 a 55.9   b 2.4
8 a 0.233   b 0.026   c 0.002   d 1.556
9 3.9
10 a >   b >   c >   d >   e <   f <
11 a 1.667, 1.286, 1.364   b $1\frac{2}{7}$, $\frac{15}{11}$, 1.49, 1.57, $\frac{5}{3}$
12 $\frac{3}{8}$, $\frac{5}{16}$, $\frac{7}{25}$, 0.205, 0.04
13 a 2.3 to 1 d.p.   b 1.3 to 1 d.p.
   c 0.3 to 1 d.p.   d 1.8 to 1 d.p.
14 a 1.3 mm   b 25.6 cm
15 30 min (0.5 is the smallest number that is 1 correct to the nearest unit.)
16 No, can be any value up to, but not including 51.5 kg.

Exam practice 6D

1 a 4, 4 tens   b 9, 9 units
   c 2, 2 hundreds   d 8, 8 tenths
   e 6, 6 hundredths   f 2, 2 tens
   g 2, 2 tens   h 5, 5 hundreds
   i 4, 4 hundreds
2 a 30   b 0.6   c 8000   d 600   e 2
   f 5   g 0.08   h 50   i 0.6
3 a 3000   b 40 000   c 70 000   d 70 000
   e 10 000   f 600   g 900   h 1000
   i 60   j 0.08   k 0.006   l 400
   m 5   n 0.4   p 10   q 0.005
   r 0.1   s 6
4 a 2   b 8   c 0.06   d 20

Exam practice 6E

1 a 60   b 70   c 200   d 20
   e 50   f 2
2 a 10 000   b 400   c 6000   d 0.04
   e 0.12   f 3
3 a 25   b 0.0016   c 0.027   d 10
4 a 0.7   b 100   c 2   d 20
5 a 4   b 20   c 0.06   d 500
6 £3000
7 a too big   b too small   c too small
   d too big   e too big   f too big
8 a B   b D   c A   d B
9 a 10   b 0.4   c 10
   d 2   e 1   f 3
10 a 0.8   b 0.07   c 500
   d 0.05   e 0.6   f 10
11 a 4   b 5   c 50
   d 7   e 0.6   f 100
12 a 60 000
   b 240 000 (overestimate as 1.8 is corrected up to 2)
13 Answer must be larger than 2.57.

Exam practice 6F

1 a 180, 173.38   b 13, 14.93   c 10, 11.43
   d 40, 37.03   e 1, 1.11   f 20, 22.71
   g 0.6, 0.52   h 160, 152.72   i 10,10.67
2 a 0.17   b 0.05   c 203.65
3 a 0.62   b 9.98   c 0.24   d 4.14
4 a 6.09   b 757.85   c 5.56   d 0.66
5 a 0.022   b 1.066   c 8.702   d 2.038
6 a 1.80   b 2.82   c 23.26   d 0.85
7 a 3.46   b 7.35   c 10.95   d 2.23
   e 0.94   f 4.86   g 0.10   h 0.02
8 84 (nearest whole number)
9 a $375 \div 0.8$ is bigger than 375 and therefore bigger than 250.
   b 469, nearest whole number.
10 £38.34 (nearest p)
11 a $\frac{1}{1-1}$
   b Denominator is 0 and you cannot divide by 0.
12 a Either put 12.5×16.2 in brackets or divide 2758 by 12.5 then divide by 16.2 (not multiply by it).
   b 14
13 a Brackets needed around $5.8 + 2.9$.
   b 2.9
14 a 5.67   b 4.90   c 2.01

Exam practice 7A

1 a 6 cm   b 7.5 cm   c 7.2 cm
   d 13.25 m   e 21.6 m   f 72 mm
2 a 200 cm   b 60 cm
3 a 500 cm   b 36 m   c 5 km   d 1500 m
4 a i 1700 cm ii 340 cm iii 900 cm iv 42 cm
   b 17 m
5 a 23 m   b 23.5 m   c 2347 cm
6 1.72 m
7 a 250 cm   b 69.3 cm   c 1200 m
   d 4.55 km   e 1.536 m   f 250 m
8 0.57m, 156 cm, 2889 mm, 3.8 m, 25 m
9 15 m
10 2.23 km
11 a 30 cm   b 7 cm   c 13 cm   d 27 cm
12 Yes, 1270 mm = 127 cm which is smaller than 135 cm.
13 $\frac{3}{40}$
14 $\frac{1}{8}$
15 216 cm

Exam practice 7B

1 a 2 ft   b 4 yds
2 a 16 km   b 64 km   c 50 miles
   d 30 cm   e 16 inches   f 44 inches
   g 50 miles   h 70 miles   i 90 miles
3 a 60 miles   b 1 mile   c 1 inch
4 1500 mm, 100 cm, 24 inches, 1 foot
5 38.1 mm
6 Yes, 84 km ≃ 52 miles.
7 402.25 km

8 a 1100 km
   b 1452 km
   c 902 (nearest mile using 1 mile ≃ 1.609 km)
9 No, 880 yds = $\frac{1}{2}$ mile
10 160 mm (6 inches ≈ 152.4 mm so you must use a wider floor board and cut it down)

Exam practice 7C

1 a 1 kg 700 g or 1.7 kg   b 850 g
   c 46.6 kg   d 52.4 kg
2 a

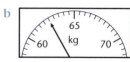

   b

3 a $\frac{1}{2}$ kg   b 1300 kg   c 0.25 t
   d 1350 g   e 45.5 t   f 0.012 t
4 a 2500 g   b 4.5 t
5 2000 g
6 0.06 t, 62 000 g, 655 kg
7 100
8 5.6 g
9 $\frac{1}{5}$
10 65.2 kg
11 $\frac{1}{10}$

Exam practice 7D

1 a 2 lb   b 24 oz
2 a 11 lb   b 1.1 lb   c 20 kg
3 12 kg
4 11 lb, 2.5 kg, 64 oz, 1500 g
5 22 lb
6 Yes, $1\frac{1}{2}$ kg = 1500 g and 2 pounds = 908 g
7 No, 100 g = $100 \div 28.4$ oz ≈ 3.5 oz. which is less than 4 oz.

Exam practice 7E

1 a 65 ml   b 185 ml
   c 5.5 litres   d 43 ml
2 a 2000 ml   b 250 ml   c 0.5 litres
   d 0.9 litres   e 1500 ml   f 1 litre
3 30 cl can
4 50 cl, 0.46 litres, 400 ml, 0.05 litres, 40 ml
5 a 1000   b $\frac{3}{10}$
6 8
7 $\frac{1}{30}$
8 16
9 6 ml
10 5
11 12
12 a 33 litres   b 22.5 litres

Exam practice 7F

1 a 2 gallons   b 12 pints
2 22.5 litres
3 261 litres

4 Less, 2 gallons ≈ 9 litres.
5 8
6 1 litre, 1 gallon, 10 pints
7 No, 10 gallons ≈ 45 litres.
8 Almost halfway between to 2 and $2\frac{1}{2}$ gallon mark since 10 litres ≈ 2.22 gallons and halfway would be 2.25 gallons.

Exam practice 7G

1 Singly 3 cost 87 compared with £1.11 in a pack.
2 6-pack, 58 p/can compared with 60 p/can.
3 100 clip pack, 31.25p/25 compared with 35 p/25.
4 Large pack, 37p/100g compared with 47 p/100g.
5 200 g pack, 65p/100g compared with 70.4p/100g.
6 a 40p   b 36.25p   c the larger jar
   d No, cost is 37p/25 g which is more expensive than the 200 g jar.
7 950 g jar
8 Buying 2 litres.

Exam practice 7H

1 a 245   b 254
2 5.55 ≤ w < 5.65 where w kg is the weight of the bag.
3 £2450
4 2.75 m
5 1.245 ≤ l < 1.255 where l mm is the diameter of the pin.
6 a 130 749   b 130 650

Exam practice 7I

1 a grams   b grams   c metres
   d km   e centimetres f litres
   g tonnes   h cm   i mm
   j cl   k litres
2 a 550 cm   b 1.5 kg   c 120 ml
   d 0.75 kg   e 4000 g   f 0.2 litres
3 a 70 inches   b 178 cm (nearest cm)
4 100 miles (≈160 km)
5 1.5 kg (≈3.3 lb)
6 80 cm
7 27 (nearest whole number)
8 17.8 (1 d.p.)
9 a 1.645 ≤ d < 1.655 where d m is the distance between the units.
   b 1651 mm is 1.651 m which is less than the maximum possible distance.
10 The sides of the cubes could be anything up to 34.5 mm but the inside could be as low as 34.15 mm.

Exam practice 8A

1 a $\frac{2}{5}$ b $\frac{4}{5}$ c $\frac{7}{10}$ d $\frac{3}{10}$ e $\frac{3}{5}$
2 a $\frac{3}{4}$ b $\frac{13}{20}$ c $\frac{9}{20}$ d $\frac{7}{20}$ e $\frac{3}{20}$
3 a $\frac{27}{100}$ b $\frac{21}{50}$ c $\frac{1}{25}$ d $\frac{17}{25}$ e $\frac{3}{50}$

4 a 0.28   b 0.81   c 0.72   d 0.12   e 0.98
5 a 0.08   b 0.01   c 0.02   d 0.06   e 0.09
6 a 0.03   b 0.67   c 0.055 d 0.125 e 0.175
7 $\frac{19}{20}$
8 $\frac{1}{10}$

Exam practice 8B

1 a 54%   b 86%   c 15%   d 39%
   e 55%   f 4%   g 1%   h 6%
   i 9%   j 2%
2 a 30%   b 20%   c 70%   d 60%
   e 10%   f 35%   g 3.5%   h 92.5%
   i 17.5% j 40.5%
3 a 132%   b 150%   c 240%   d 105%
   e 255.5%
4 a 50%   b 75%   c 50%   d 60%
   e 5%
5 a 150%   b 110%   c 120%   d 225%
   e 175%
6

Fraction	Percentage	Decimal
$\frac{1}{5}$	20%	0.20
$\frac{1}{4}$	25%	0.25
$\frac{1}{2}$	50%	0.5
$\frac{7}{20}$	35%	0.35
$\frac{4}{5}$	80%	0.8
$\frac{1}{8}$	12.5%	0.125
$1\frac{1}{5}$	120%	1.2

7 a 45%   b 0.48   c $\frac{9}{25}$
   d $\frac{14}{25}$ e 0.32   f 84%
8 40%
9 85%

Exam practice 8C

1 a 8.3%   b 55.6%   c 13.3%   d 42.9%
   e 66.7%   f 16.7%   g 133.3% h 122.2%
2 42%
3 33%
4 a $\frac{15}{52}$ b 29%
5 0.175
6 No, $\frac{1}{3} = 33\frac{1}{3}\%$
7 a 0.6   b $\frac{1}{8}$
8 $\frac{1}{10}$, 11%, 0.12, 21%

Exam practice 8D

1 a £20   b 3 p   c £1.20
   d £43   e £24   f £3
2 a £18   b £500   c £450
   d 15 kg e 200 g f 30 m
3 a £1   b 12p   c 0.84 km
4 39p

5 £5000
6 £4.23 nearest p
7 180
8 £4.62
9 200
10 £290
11 1530
12 £8.91

Exam practice 8E

1 a £232.40   b £20.75   c £2.49
    d £287.88   e £56.70   f £25.15
    g 166.6 g   h 3.5 cm   i 149.5
    j £44   k 67.0 miles
    l 6.5 min (all to 1 d.p.)
2 127.5 sq m
3 £15.73 (nearest p)
4 £4410
5 £437.50
6 £1113
7 a 12 346 000 (nearest 1000)
    b 5 056 000 (nearest 1000)
8 a £18.20   b £122.20
9 a £20.75   b £809.25
10 249
11 It is 30% of the total cost.

Exam practice 8F

1 a £15   b £12.50   c £14   d £85
2 a £32   b £450   c £500   d £337.50
3 £112.50
4 £31 650
5 54 p
6 £1000
7 £750
8 £2000
9 a £80   b 8%
10 a £616   b £77

Exam practice 8G

1 a 3%   b 45%   c 10%   d 10%
2 a 60%   b 25%   c 112.5% d 10%
    e 12%   f 33%   g 40%
3 30%
4 26% (nearest whole number)
5 30%
6 112.5%
7 7.8% (1 d.p.)
8 24%
9 109% (nearest whole number)
10 $3\frac{1}{2}$%
11 a £425   b 34%
12 a 52%   b 48%
13 62.5%
14 a 8%
    b 10%
    c Yes, 25 out of 275 is 9.09% which is less
        than 10%.

Exam practice 8H

1 7.8 kg
2 £30.60
3 £25
4 44
5 5%
6 a £35   b 12.5%
7 33
8 90
9 £845
10 £8.40
11 £999 (nearest whole number)
12 a £36   b £8.50
13 31
14 20%
15 $33\frac{1}{3}$%
16 £4.20
17 12%
18 29% (nearest whole number)
19 2.52 kg
20 £3840
21 a £170   b £435

Exam practice 9A

1 a 2:3   b 1:3   c 1:4
    d 2:5   e 5:1   f 25:9
2 a 1:4   b 12:5   c 4:1
    d 1:3   e 5:2   f 5:1
3 a 2:1   b 1:2   c 1:2
    d 2:1   e 2:3   f 3:5
4 a 16:3   b 1:6   c 2:1
    d 5:1   e 4:1   f 40:1
5 b and d
6 Yes, multiplying each number in the ratio
    25:15 by 2 gives 50:30.
7 No, $\frac{1}{2}:\frac{1}{4} = 2:1$.
8 3
9 a 1:40   b 11:40   c 3:8
10 3:4
11 5:3
12 5:8
13 1:2
14 1:300 000
15 a 3:2   b 2:5
16 1:200

Exam practice 9B

1 100 m
2 1:25 000
3 2 km
4 a 100 m   b 1.2 km
5 4.5 km
6 1:25 000
7 a 2 cm   b 2 km
8 1: 2 000 000
9 a $\frac{1}{2}$ km   b 5 km   c 20 cm
10 a 5 cm   b 1:90
11 24 m
12 a 10 m   b 40 m

13 a 1.08 m   b 10 cm
14 3:2
15 a teeth on little wheel: teeth on large wheel is 1:4
   b 4
   c 2
16 a 5   b 8
17 a $4\frac{1}{2}$ b 4
18 a 4:3   b 3:2   c C makes two turns.
19 30 cm
20 25 cm

Exam practice 9C

1 £10.80
2 £7.92
3 £1.20
4 $4\frac{1}{3}$ km
5 £8.30
6 0.75 units
7 a 1.5p   b 40 min
8 a 12 sq m   b 60 sq m
9 a £5.50   b £198

Exam practice 9D

1 3 hr
2 a 12.5   b 3.6 hr (3 hr 36 min)
3 £61.20
4 £265.65
5 500
6 66
7 £324
8 £18.90
9 £170
10 a 345 g   b 135 ml
11 a 12 oz   b 4.5 oz
12 a 150 g   b 60 g
13 30 g flour, 30 g margarine, 150 g cheese, 450 ml milk
14 £5.265, £5.27

Exam practice 9E

1 a 48p:32p   b 12 cm:20 cm
   c £20:£25   d £2.50:£17.50
2 15:25
3 30p:45p
4 16
5 £294
6 £8, £16
7 48 g
8 1.8 m
9 a 4 litres   b 21 litres
10 a 16 kg, 24 kg, 40 kg   b £8, £16, £16
11 120 g
12 £165 for Pete and Zoe, £220 for Sara
13 100 g, 160 g
14 £2.40, £3.30
15 36
16 0.392 litres (3 s.f.)
17 1 litre

18 a 252 sq m   b 105 sq m
19 a 0.0166 litres
   b 0.00997 litres (both to 3 s.f.)
20 0.6 kg
21 $\frac{2}{9}$ litre
22 Jo £80 645, Andy £56 452, Greg £112 903

Exam practice 10A

1 a Sept   b Thursday   c i 14th   ii 23rd
2 9th May
3 a 27th August   b 49   c 4
4 a Donna   b Rosemary   c 2025
5 a 17   b D S Short   c 2012
6 a 2 hr 50 min   b 20 days 10 hr
   c 126 sec   d 150 min
7 a $\frac{2}{3}$ b $\frac{1}{100}$
8 a 10.25 a.m.   b 11.10 a.m.   c 45 min
9 a 17 min   b 1625
10 a 10 min   b 2 hr 20 min
11 a 1 hr 45 min   b 8 hr 40 min
   c 2 hr 10 min
12 7.05 p.m.
13 2 hr 20 min
14 a 25 min   b 8.10 p.m.
15 14 hr 58 min
16 a 6 hr 30 min   b 8 hr 29 min
   c 8 hr   d 3 hr 5 min
17 1915
18 1905
19 a 1st 2 hr 22 min, 2nd 2 hr 17 min
   b Bewley and Peters Farm, takes shortest time.
20 a 19.43   b 47 min
21 a 2 hr 35 min   b 15.12
22 £34.40
23 3 hr 5 min
24 41 min 40 sec
25 a 9.47   b 12.47
26 a £6   b No. Runs out at 4.47.

Exam practice 10B

1 a 12°C   b 28°F   c 28°F (32°F = 0°C)
2 a 105°F   b 40°C   c 20°C   d 68°F
3 a 86°F   b 14°F   c 27°C ($26\frac{2}{3}$ °C)
   d −12°C (−12.22…°C)
4 $38\frac{1}{3}$°C (38.33…°C)
5 113°F
6 121°C (121.11…°C)

Exam practice 10C

1 15 km/h
2 112.5 m.p.h.
3 4 km/h
4 50 m.p.h.
5 52 m.p.h.
6 27 km/h
7 6 m.p.h.
8 a $3\frac{3}{4}$ m.p.h.   b 40 m.p.h.
9 45m/sec
10 5.2 m.p.h

11 20 m.p.h.
12 $1\frac{1}{2}$ hr
13 a 109 km/h
    b 95 km/h (both to nearest whole number)
14 113 km/h (nearest whole number)
15 a 57 m.p.h
    b 56 km/h
    c 57 m.p.h. (a mile is further than a kilometre)

Exam practice 10D

1 a €85 (nearest whole number)
   b 536 lira
   c $270
2 $340
3 $1380
4 £192.31
5 £316.46
6 80 000 yen
7 90 p
8 £27.03
9 a £192.31   b £15.38
10 a €18   b £80
11 £1 = 61.6 rupees
12 a €39   b £13   c £23
13 London
14 a £1 = €1.52
   b €1 = 61 p or £1 = €1.64
15 a $127.50   b £21.05
16 Florida

Examination practice paper

Section A

1 2 bars cost 80p at Lo-prices, but 90p at Costsavers.
2 a i Any of 17, 3, 9 or 41   ii 9   iii 24
   b 1, 21, 3, 7
3 a £41   b 6
4 a 539   b 48
   c Half of 103 is 51.5 and there can't be a half boy.
5 a 2.89   b 7
   c i 2.3780821   ii 2.38
6 a 50 minutes
   b 2 hours 5 minutes
7 a £8.67   b 30%
8 a 24 550   b 24 649

Section B

1 a 0.5   b $\frac{1}{5}$ c 0.6
2 £2.60
3 a 1001, 1010, 1100
   b 0.075, 0.39, 0.6
   c −3, −2, 1
4 a 214   b 34   c 36
   d 12   e 11 092
5 Any number with more than 1 decimal place.
6 a 0.08   b 8   c $\frac{5}{8}$
7 36 m.p.h.
8 5
9 40 litres
10 484

Index

A

addition
   decimals 64
   fractions 52
   mixed numbers 53
   negative numbers 23–4
   whole numbers 3, 4
approximations
   by rounding 82–3
   summary of key points 87
arithmetic rules
   addition 23–4
   decimal multiplication 67–8
   division 26
   multiplication 26
   subtraction 23–4
average speed 140

B

base 37–8
'best buys' 98
'best value for money' 98
bills 69
brackets 19
Brahmagupta 45

C

calculator
   percentages 111
   reducing answers displayed 84–5
calendar 134
capacity
   imperial units 97
   metric units 95
Celsius 138
common factors 33
conversions
   money 144
   speed 140
   temperature 138
   units of capacity 97
   units of length 91
   units of mass 94
counting numbers 29
cube numbers 38
cube root 40
currencies 142–3

D

decimals
   addition 64
   bills 69
   comparing 62
   division 67–8
   division by 10, 100, 1000, ... 66
   fractions 63
   inexact answers 108
   money 69
   multiplication 67–8
   multiplication by 10, 100, 1000, ... 66
   percentages 106–7
   recurring 71
   rounding 77–8
   subtraction 64
   summary of key points 74
decimals places 61–2
denominator 45, 105
digit 1
directly proportional 126
division
   by 10, 100, 1000, ... 16
   by a fraction 56
   by whole numbers less than 10: 14
   decimals 67–8
   decimals by 10, 100, 1000, ... 66
   directed numbers 26
   exact 29
   in a given ratio 128–9
   index notation 41
   long 17
   remainders 15
dollar 142

E

equivalent fractions 47
estimating
   answers to calculations 82–3
   by rounding 82–3
   summary of key points 87
even numbers 29
exchange rates 142–3

F

factors
   basics 31
   common 33

highest common 33–4
   prime 31, 42–3
Fahrenheit 138
fractions
   addition 52
   basics 45
   comparing 48
   decimals 71
   equivalent 47
   improper 49–50
   lowest possible terms 47
   mixed numbers 49–50
   multiplication 54
   of a quantity 57
   percentages 106–7
   proper 49–50
   quantity as fraction of another 50–1
   ratio 128
   reciprocals 55–6
   simplifying 47
   subtraction 52
   summary of key points 60

G
gear ratio 123

H
highest common factor 33–4, 42–3

I
imperial units
   capacity 97
   length 90–1
   mass 94
improper fractions 49–50
   multiplication 55
inexact answers
   decimals 108
   measurements 101–2
integers 29
interest 112–13
interest rate 112–13

L
least common multiple 35–6
length
   imperial units 90–1
   metric units 88–9
long division 17
long multiplication 13
lowest common multiple 42–3

M
magic square 10
map ratio 122
mass

imperial units 94
metric units 92
measures
   always rounded 101–2
   summaries of key points 104, 146
metric units
   capacity 95
   length 88–9
   mass 92
mixed calculations
   addition and subtraction 8
   order of operations 18–19
mixed numbers 49–50
   addition 53
   multiplication 55
   subtraction 53
money 69, 142–3
multiples
   basics 32
   common 35
   least common 35–6
multiplication
   by 10, 100, 1000, … 11–12
   by whole numbers less than 10: 10
   decimals 67–8
   decimals by 10, 100, 1000, … 66
   directed numbers 26
   fractions 54
   improper fractions 55
   index notation 41
   long 13
   mixed numbers 55

N
natural numbers 29
negative numbers 21
   summary of key points 28
notation
   approximately equal to 83, 91
   cube root 40
   decimal place 78
   highest common factor 33
   index 37–8
   least common multiple 35
   per annum 113
   power 37–8
   ratio 120
   significant figures 82
   square root 39
   standard form 72–3
number line 6, 21
numbers
   base 37–8
   counting 29
   cube root 40
   cubed 38

digits 1
even 29
integers 29
mixed 49–50
natural 29
negative 21
odd 29
perfect square 39–40
positive 21
power 37–8
prime 29, 31
square 29, 84
square root 39–40, 84
squared 38
types 29
summary of key points 44
numerator 45

O
odd numbers 29

P
paper sizes 131
'per cent' 105
percentages
    calculator 111
    decimals 106
    decrease 116
    fractions 105
    increase 116
    interest 112–13
    interest rate 112–13
    of a quantity 109–10
    quantity as percentage of another 114
    simple interest 112–13
    summary of key points 119
    Value Added Tax 109, 110, 111
perfect square 39–40
perimeter 4
π (pi) 73
place value 1, 61
positive numbers 21
power of a number 37–8
prime factors 31, 42–3
prime numbers 29, 31
profit 117
proper fractions 49–50
proportion 125–7
    summary of key points 132

R
ratio
    basics 120
    division in a given ratio 128–9
    fractions 128
    simplifying 120–1
    summary of key points 132
    use of 122–3
reciprocals 55–6
recurring decimals 71
remainders 15
rounding
    answers 79
    degree of accuracy 79
    for estimating answers 82–3
    range of rounded number 100–1
    to nearest 1, 10, 100 ... 75–6
    to number of decimal places 77–8
    to significant figures 82

S
scale drawings 122
significant figures 81–2
simple interest 112–13
simplifying fractions 47
simplifying ratio 120–1
speed 140
square numbers 29, 38, 84
square root 39–40, 84
standard form 72–3
subtraction
    decimals 64
    fractions 52
    mixed numbers 53
    negative numbers 23–4
    whole numbers 6–7

T
temperature 138
thermometer 21
time 133–4
timetables 136–7, 142
triangular numbers 43
triangular pattern of numbers 25
12-hour clock 134
24-hour clock 134

V
Value Added Tax 109, 110, 111

W
word problems 9